D1163781

CHICAGO PUBLIC LIBRARY
HAROLD WASHINGTON LIBRARY CENTER
R0021984405

J. VAN ALPHEN

# RUBBER CHEMICALS

## REVISED AND ENLARGED EDITION

*Edited by*

C. M. VAN TURNHOUT

*Plastics and Rubber Research Institute TNO, Delft, Holland*

D. REIDEL PUBLISHING COMPANY

DORDRECHT-HOLLAND / BOSTON-U.S.A.

This revised and enlarged edition of J. van Alphen's **Rubber Chemicals**
is published in collaboration
with the Plastics and Rubber Research Institute TNO, Delft, Holland

The Plastics and Rubber Institute TNO is grateful
to Shell International Chemical Company Limited
for assistance in the production of this book



Library of Congress Catalog Card Number 73-81826

ISBN 90 277 0349 3

Sold and distributed in the U.S.A., Canada, and Mexico
by D. Reidel Publishing Company, Inc.
306 Dartmouth Street, Boston,
Mass. 02116, U.S.A.

Sold and distributed in all other countries
by D. Reidel Publishing Company,
P.O. Box 17, Dordrecht, Holland





Printed in Holland by B.V. Drukkerij De Eendracht, Schiedam

# Preface



The ever increasing number of rubber chemicals that are being introduced has led to an increasing rate of depletion of the first edition of Dr. J. van Alphen's book which so usefully provided chemical names, trade names and the names of suppliers of a very large number of commonly used rubber additives. The need for a revised edition of this valuable book, original edited by „Rubber-Stichting" in 1956 was recognized by Shell International Chemical Company who was ready and willing to sponsor the necessary work to update and extend the contents of the book, preparations of which — following the death of Dr. van Alphen — were already taken in hand by Mr. R. J. Kuipers, at that time a TNO rubber technologist. Unfortunately Mr. Kuipers did not live to see the completion of this work and the Economic Technical Information Centre of the TNO Plastics and Rubber Research Institute was prepared to undertake the extensive task of cataloguing the expanded range of rubber chemicals in the following groups, largely following the original pattern set by Dr. van Alphen. The present book contains the following chapters:

peptizing agents
vulcanizing agents
accelerators
activators
retarders
blowing agents
antidegradants
co-agents

It is gratefully acknowledged that several producers have responded to the request for data and information, so that all chapters are now fully up to date.
The chapter on peptizers now includes newly developed products that are already active at comparatively low temperatures as well as materials displaying prolonged peptizing activity during compounding. With regard to vulcanizing agents relatively few developments have taken place and this chapter has not been changed substantially.

It goes without saying that the important groups of vulcanization accelerators and activators have been treated with considerable care.

Specific attention has been paid to include those accelerators which show different degrees of 'delayed action', so important in modern compounding, and those materials which consist of a combination of accelerators with other active ingredients resulting in special effects. The chapter on retarders reflects the comparative few developments in this field, but the number of blowing agents has considerably increased over the years, particularly because of the need for higher decomposition temperatures, especially with a view to injection moulding. Consequently this chapter has also been expanded significantly.

A new chapter has been introduced on antidegradants. They constitute a wide field of additives including antioxidants, anti-ozonants, anti-flex cracking agents, heat stabilizers and metal deactivators. This chapter has thus been significantly extended in comparison with the antioxidant section in the first edition.

In the last decade carbon black manufacturers have adopted considerable standardization with respect to types and grades of black. The corresponding rationalization has led to the envisaged simplification of the situation and carbon black manufacturers nowadays supply extensive and comprehensive data on the types now available. It has consequently been decided to omit the section on carbon black. Emulsifiers that were included in the first edition have now been left out as it was felt that ready reference should be provided to materials of general use throughout the rubber industry.

The single index on trade marks and chemical names of compounds has been replaced by two separate alphabetical indices. The original edition was in English and German, but in view of the general use of the English language in many areas of technical endeavour it was decided to issue the present edition in English only.

I trust that the book in its new form will prove to be even more useful than its predecessor, especially to technologists anywhere in the large rubber processing industry.

J. M. Goppel
Director of Research
KONINKLIJKE/SHELL PLASTICS LABORATORIUM DELFT

# Abbreviations

## Rubber types

| | |
|---|---|
| Acrylic rubber | ACM |
| Butadiene rubber | BR |
| Butyl rubber | IIR |
| Chloroprene rubber | CR |
| Chlorosulfonated polyethylene | CSM |
| Ethylene-propylene rubber | EPM |
| Ethylene-propylene-diene terpolymer | EPDM, EPT |
| Isoprene rubber | IR |
| Natural rubber | NR |
| Nitrile-butadiene rubber | NBR |
| Silicone rubber | Si, PSi, CSi, PVSi |
| Styrene-butadiene rubber | SBR |
| Polyacrylate rubber | ANM |
| Polysulphide rubber | none |

## Rubber chemicals

| | |
|---|---|
| Accelerators | Acc. |
| Activators | Act. |
| Antidegradants | Antidegr. |
| Antioxidants | Oxi. |
| Antiozonants | Ozo. |
| Flex-cracking agents | Flex. |
| Heat-stabilizers | Heat. |
| Metal-Poison inhibitors | Inh. |
| Blowing agents | Ba. |
| Co-agents | Ca. |
| Peptizers | Pept. |
| Retarders | Ret. |
| Vulcanizing agents | Vulc. |

## Miscellaneous

| | |
|---|---|
| Food and Drug Administration approved | FDA appr. |
| Freezing Point | F.P. |
| Melting Point | M.P. |
| Melting Range | M.R. |
| Molecular Weight | M.W. |
| Specific Gravity | S.G. |
| No longer on selling range | † |
| See also | △ |
| Delayed action | Del. act. |

# Names of Manufacturers

| Abbreviations | Full addresses |
|---|---|
| Aceto | **Aceto Chemical Company Inc.**<br>126-02 Northern Blvd., Flushing,<br>New York 11368, U.S.A. |
| ACNA Montecatini | **A.C.N.A. Montecatini Edison**<br>Largo Donegani 1-2, Milano, Italy |
| Akron | **Akron Chemical Company**<br>255 Fountain Street, Akron, Ohio 44304, U.S.A. |
| Alkali | **Alkali Chemical Corporation India Ltd.**<br>Rishra, West Bengal, India |
| Allied | **Allied Chemical Corporation, Plastics Div.**<br>P.O. Box 365, Morristown, New Jersey 07960,<br>U.S.A. |
| Cyanamid | **American Cyanamid Company,**<br>**Rubber Chemicals Dept.**<br>Bound Brook, New Jersey 08805, U.S.A. |
| American Hoechst | **American Hoechst Corp.**<br>4331 Chesa Peake Drive, Charlotte,<br>N.C. 28208, U.S.A. |
| Anchor | **Anchor Chemical Co. Ltd.**<br>Clayton, Manchester M 11 4 SR, England |
| Arwal | **Arwal Chemicals Inc.**<br>1961 West Market Street, P.O. Box 429, Akron,<br>Ohio 44309, U.S.A. |
| Ashland | **Ashland Chemical Comp.**<br>P.O. Box 1503, Houston, Texas 77001, U.S.A. |
| BASF | **Badische Anilin & Soda Fabrik A.G.**<br>6700 Ludwigshafen/Rhein, Germany |
| Bayer | **Bayer A.G.**<br>509 Leverkusen, Germany |
| Bennett | **D. G. Bennett Chemicals**<br>11A St. John's Hill, London SW 11 . 1TS, England |
| Bozzetto | **Bozzetto Industrie Chimiche S.p.A.**<br>P.O. Box 240, Pedrengo (Bergamo), Italy |

| Abbreviations | Full addresses |
|---|---|
| CdF | **CdF Chimie**<br>Cedex No. 5, Tour Aurore, 92 Paris, France |
| Chemko | **Chemko**<br>Strázske, Czechoslovakia |
| Ciba Geigy | **Ciba Geigy A.G.**<br>CH-4000 Basel 7 (Rosenthal), Switzerland |
| Conestoga | **Conestoga Chem. Corp., Div. of Chemetron**<br>Wilmington Ind. Park, Foot of East 7th Str.<br>Wilmington, Delaware 19801, U.S.A. |
| Degussa | **Degussa**<br>P.O. Box 3993, 6000 Frankfurt/Main 1, Germany |
| Dimitrova | **Chemické závody Juraja Dimitrova**<br>Bratislave, Czechoslovakia |
| Du Pont | **E.I. Du Pont de Nemours & Co. Inc.**<br>Wilmington, Delaware 19898, U.S.A. |
| Durham | **The Durham Chemical Group**<br>Birtley, County Durham, England |
| Eastman | **Eastman Kodak Company,**<br>**Tennessee Eastman Comp. Div.**<br>Kingsport, Tennessee 37662, U.S.A. |
| Fairmount | **Fairmount Chemical Co. Inc.**<br>117 Blanchard Street, Newark,<br>New Jersey 07105, U.S.A. |
| Fisons | **Fisons Industrial Chemicals Ltd.**<br>Willows Works, Derby Road, Loughborough,<br>Leicestershire, England |
| Goodrich | **B. F. Goodrich Chem. Comp.**<br>3135 Euclid Avenue, Cleveland,<br>Ohio 44115, U.S.A. |
| Goodyear | **The Goodyear Tire and Rubber Company**<br>Akron, Ohio 44316, U.S.A. |
| Grandel | **Deutsche Oelfabrik Dr. Grandel & Co.**<br>P.O. Box 111929, 2 Hamburg 11, Germany |
| Hall | **The C. P. Hall Company**<br>4460 Hudson Drive, Stow, Ohio 44224, U.S.A. |

| Abbreviations | Full addresses |
|---|---|
| Hasselt | **Van Hasselt N.V.**<br>Amsterdamseweg 18, Amersfoort,<br>The Netherlands |
| Hercules | **Hercules Inc. Intern. Dept.**<br>Wilmington, Delaware 19899, U.S.A. |
| ICI America | **ICI America Inc.**<br>Stamford, Connecticut, U.S.A. |
| Icianz | **Icianz Ltd.**<br>Melbourne 3001, Australia |
| ICI (India) | **ICI (India) Pty. Ltd.**<br>Chowringhee, Calcutta, India |
| ICI | **Imperial Chemical Industries Ltd.**<br>Millbank, London SWIP 3JF, England |
| Ticino | **Industria Chimica del Ticino S.p.A.**<br>Via del Porto, 28040 Marano Ticino, Italy |
| Kenrich | **Kenrich Petrochemicals Inc.**<br>Foot of East 22nd Street, Bayonne,<br>New Jersey 07002, U.S.A. |
| Lucidol | **Lucidol, Div. of Wallace & Tiernan Inc.**<br>1740 Military Road, Buffalo,<br>New York 14240, U.S.A. |
| May & Baker | **May & Baker Ltd.**<br>Dagenham, Essex RM 10 7 XS, England |
| Merck | **Merck & Co. Inc.**<br>Rahway, New Jersey 07651, U.S.A. |
| Metallgesellschaft | **Metallgesellschaft A.G., Abt. TA**<br>Reuterweg 14, 6000 Frankfurt/Main, Germany |
| Monsanto | **Monsanto Corp.**<br>800 N. Lindbergh Blvd.<br>St. Louis, Missouri 63166, U.S.A. |
| Nat. Polychemicals | **National Polychemicals Inc.**<br>Wilmington, Massachusetts 01887, U.S.A. |
| Naugatuck S.p.A. | **Naugatuck S.p.A.**<br>Corso Vinzaglio 35, 10121 Torino, Italy |
| Neville | **Neville Chemical Comp.**<br>Neville Island, Pittsburgh, Pa. 15225, U.S.A. |

| Abbreviations | Full addresses |
| --- | --- |
| Neville Synthèse | **Neville Synthèse Organics Inc.**<br>Neville Island, Pittsburgh, Pa. 15225, U.S.A. |
| Newalls | **Newalls Insulation and Chemical Co. Ltd.**<br>Washington, Co. Durham, England |
| Norac | **The Norac Co. Inc.**<br>405 South Motor Avenue, P.O. Box F, Azusa, California 91703, U.S.A. |
| Nourychem | **Nourychem Corp.**<br>2153 Lockport-Olcott Road, Burt, New York 14028, U.S.A. |
| Noury | **Noury & v. d. Lande N.V. ***<br>P.O. Box 10, Deventer, The Netherlands |
| Ouchi Shinko | **Ouchi Shinko Chemical Industrial Co. Ltd.**<br>3-7, 1-chome, Kobuna-cho, Nihombashi, Chuo-ku, Tokyo, Japan |
| Pearson | **William Pearson Ltd.**<br>Clough Road, Bull, Yorkshire, England |
| Pennwalt | **Pennwalt Corp.**<br>900 First Avenue, King of Prussia, Pa. 19406, U.S.A. |
| Pitt Consol | **Pitt Consol Chemical Company**<br>380 North Street, Teterboro, New Jersey 07608, U.S.A. |
| Polychimie | **Polychimie**<br>Le Pressoir Vert, 45 Semoy, (Orléans), France |
| Prochim | **Prochim**<br>59 Courchelettes, P.O. Box 236, Douai, France |
| Raschig | **Dr. F. Raschig GmbH**<br>6700 Ludwigshafen/Rhein, Germany |
| Reagens | **Reagens, Societá per Azioni, Industria Chimica**<br>S. Giorgio di Piano, Bologna, Italy |
| Reichhold Albert | **Reichhold Albert Chemie A.G.**<br>Harvestehuderweg 18, 2 Hamburg 13, Germany |

* Change of name: **AKZO CHEMIE B.V.**
Sales office Deventer

| Abbreviations | Full addresses |
|---|---|
| Reichhold | **Reichhold Chemicals Inc.**<br>RCI Building, White Plains,<br>New York 10602, U.S.A. |
| Rhein-Chemie | **Rhein-Chemie Rheinau GmbH**<br>P.O. Box 84 and 104, 6800 Mannheim 81,<br>Germany |
| Rhône Poulenc | **Société des Usines Chimiques Rhône Poulenc**<br>22, Avenue Montaigne, 75 Paris VIIIe, France |
| Robinson | **Robinson Brothers Ltd.**<br>Ryders Green, West Bromwich, Staffordshire,<br>England |
| Ross | **Frank B. Ross Co. Inc.**<br>6-10 Ash Street, Jersey City,<br>New Jersey 07304, U.S.A. |
| Rubber Regenerating | **The Rubber Regenerating Comp. Ltd.**<br>Trafford Park, Manchester M 17 1 DT, England |
| Sanshin | **Sanshin Chemical Industry Co. Ltd.**<br>Yanai Minato, Yanai City, Yamaguchi Pref., Japan |
| Sartomer | **Sartomer Resins Inc.**<br>P.O. Box 56, Essington, Pa. 19029, U.S.A. |
| Schill & Seilacher | **Schill & Seilacher Chem. Fabrik**<br>Moorfleeterstrasse 28, P.O. Box 460,<br>2000 Hamburg 74, Germany |
| Schlickum | **Schlickum Werke J. Schlickum & Co.**<br>P.O. Box 225, 2000 Hamburg 36, Germany |
| Shell | **Shell International Chemical Company Ltd.**<br>Shell Centre, London S.E.1. 7PG, England |
| Organo Synthèse | **Société Française d'Organo Synthèse S.A.**<br>159, Avenue de Roule, 92 Neuilly/Seine, France |
| Spolek | **Spolek Pro Chemikou A Hutni Vyrobu**<br>Ustí nad Labem, Czechoslovakia |
| Stauffer | **Stauffer Chemical Company**<br>380 Madison Avenue, New York,<br>New York 10017, U.S.A. |
| Sumitomo | **Sumitomo Chemical Co. Ltd.**<br>15, 5-chome, Kitahama, Higashi-ku, Osaka, Japan |

| Abbreviations | Full addresses |
|---|---|
| Uniroyal | **Uniroyal Inc.**<br>Spencer Street, Naugatuck,<br>Connecticut 06770, U.S.A. |
| UOP | **Universal Oil Products Company**<br>State Highway 17, East Rutherford,<br>New Jersey 07073, U.S.A. |
| Vanderbilt | **R. T. Vanderbilt Company Inc.**<br>230 Park Avenue, New York,<br>New York 10017, U.S.A. |
| Vondelingenplaat | **Fabriek van Chemische Producten Vondelingenplaat N.V.**<br>P.O. Box 7120, Rotterdam, The Netherlands |
| Vychodoceské | **Vychodoceské Chemické závody**<br>Kolin, Czechoslovakia |
| Wallace & Tiernan | **Wallace & Tiernan Chemie GmbH**<br>P.O. Box 149, 887 Günzburg/Donau, Germany |

# Accelerators

## INORGANIC

**1**

1. Lead oxides
2. —
3. Mix Lpb 80 — Bozzetto a
   Mix Pb 80 — Bozzetto b
   Polyminium — Polychimie b
   Polytharge grades A, B, D — Polychimie a
   RC Granulat PbO — Rhein Chemie a
   RC Granulat $Pb_3O_4$ — Rhein Chemie b
   — — Anchor c
4. a. PbO dispersion. Acc. for CR; act. for NR, NBR, SBR, IIR; vulc. for CR and CSM.
   b. $Pb_3O_4$ dispersion. Acc. for CR; act. for IIR,
   c. PbO powder (litharge). Vulc. for CR.
   △ Act. 1
   △ Vulc. 1

## ORGANIC

### Aldehyde-ammonia compounds

**2**

1. Hexamethylenetetramine (with additives)
2. $(CH_2)_6N_4$
3. Aceto HMT — Aceto
   Eveite UR — ACNA Montecatini
   Herax UTS — Dimitrova
   Hexa K — Degussa
   Hexalit grades S,01,02 — Chemko
   Nocceler H — Ouchi Shinko
   RC Granulat Hexa (80% hexa and 20% saturated hydrocarbons and dispersion agents) — Rhein Chemie

Rhenocure Hexa — Rhein Chemie †
Sanceler H — Sanshin
Soxinol H — Sumitomo
Vulkacit H 30 — Bayer

4. Hygroscopic white powder. M.W. 140.19. Sublimates at 263 °C. S.G. 1.3.
Acc. for NR, SBR, NBR; act. for mercapto-, sulphenamide-, thiuram- and zinc dithiocarbamate-type acc.
△ Act. 15

**3**
1. Ethylchloride - formaldehyde - ammonia reaction product
2. —
3. Trimene Base — Uniroyal
Trimene (Trimene Base with stearic acid 1 : 1) — Uniroyal a
Vulcafor EFA — ICI
4. Dark brown viscous liquid. S.G. 1.09.
a. Thick paste, S.G. 0.97.
Acc. and latex foam stiffener for NR, SBR.

## Aliphatic amines and their condensation products

**4**
1. Aldehyde-amine condensation product
2. —
3. Vulcafor PT — ICI †
4. Brown resinous mass. S.G. 1.05.

**5**
1. Alkylamine
2. —
3. Rhenocure PA — Rhein Chemie
4. Light coloured viscous liquid. B.R. 215–225 °C.; S.G. 0.93.
Semi-ultra acc. Crosslinking agent for acrylate elastomers.

**6**
1. Butyraldehyde - monobutylamine condensation product
2. —
3. Accelerator 833 — Du Pont

4. Reddish amber liquid. F.P. 116 °C.; S.G. 0.86.
   Staining acc. for NR, SBR and reclaim. Also for latex. Acc. for self-curing CR cements.

**7**
1. Cyclohexyl ethyl amine
2. $\overline{(CH_2)_5}CH.NH.C_2H_5$
3. Vulkacit HX — Bayer
4. Pale yellow liquid. M.W. 127; B.P. 165 °C.; S.G. 0.85.
   Sec. acc. and act.
   △ Act. 12

**8**
1. Dibutylamine
2. $(C_4H_9)_2NH$
3. — — B.A.S.F.
4. Clear liquid. S.G. 0.75.
   Sec. acc. for Vulkacit P extra N (Bayer).

**9**
1. Polyethylene-polyamine
2. —
3. Vulkacit TR — Bayer
4. Yellow to reddish-brown liquid. S.G. 0.99.
   Semi-ultra acc. used alone or in combination with Vulkacit 1000.

**10**
1. Tricrotonylidenetetramine (with dispersion agents)
2. —
3. Vulkacit CT-N — Bayer
4. Viscous brown oil. S.G. 1.05.
   Acc. for NR, IR, BR, SBR, NBR.

**Aromatic amines and their condensation products**

**11**
1. Acetaldehyde-aniline condensation products
2. —
3. Crylene (mixture with stearic acid 67 : 33) — Uniroyal a
   Nocceler K — Ouchi Shinko b
   VGB — Uniroyal c
   Vulcafor RN — ICI †

4. a. Thick brown paste. S.G. 1.014.
   b. Reddish-brown powder. M.P. 55 °C.
   c. Brown resinous powder. M.R. 60–80 °C.; S.G. 1.15.
   Staining acc. for NR; oxi, heat for NR.
   △ Antidegr. 40

**12**
1. Anhydroformaldehyde - aniline condensation product
2. —
3. Vulcafor MA — ICI †
   Vulcafor MA — ICI (India) †
4. Yellow powder. M.P. 133 °C.

**13**
1. Anhydroformaldehyde p-toluidine condensation product
2. —
3. Vulcafor MT — ICI †
4. White powder. M.P. 133 °C.

**14**
1. Butyraldehyde - aniline condensation product
2. —
3. Accelerator 21 — Anchor
   Antox Special — Du Pont
   Beutene — Uniroyal
   Butanyl-1 — Ticino
   Nocceler 8 — Ouchi Shinko
   Rapid Accelerator 300A — Rhône Poulenc
   Vulcafor BA — ICI
4. Orange-red oily liquid. S.G. 0.94–1.02.
   Semi-ultra acc. for Neoprene W and hydrocarbon rubber; act. for thiazoles, thiurams and guanidines; antidegr. for CR.
   △ Act. 11
   △ Antidegr. 41

**15**
1. 4,4'-Diamino diphenyl methane
2. $CH_2(C_6H_4.NH_2)_2$
3. Robac 4.4 — Robinson a
   Tonox — Uniroyal b
4. a. Light brown powder. M.W. 198.3; M.R. 75–85 °C.
   Acc. for CR; ret. for IIR; anti-frosting agent for NR.
   b. Brown waxy lump. S.G. 1.18.
   △ Antidegr. 42
   △ Ret. 2

**16**
1. N,N-Di-(N'-ethylidene-anilino) aminobenzene
2. $C_6H_5-NH-CH(CH_3)-N(C_6H_5)-CH(CH_3)-NH-C_6H_5$
3. Eveite A — ACNA Montecatini
4. Brown oily liquid. M.W. 331.45; S.G. 1.06–1.07.
   Acc. and antidegr.
   △ Antidegr. 43

**17**
1. Heptaldehyde - aniline condensation product
2. —
3. Heptene Base — Uniroyal
4. Dark brown liquid. S.G. 0.92.
   Acc. for NR; act. for thiazole- and thiuram-type acc.
   △ Act. 14

**18**
1. Butyraldehyde - butylidene-aniline reaction product
2. —
3. A-32 — Monsanto a
   Eveite 101 — ACNA Montecatini b
4. a. Dark-amber liquid. S.G. 1.01.
   Used alone or in combination with thiazoles or thiurams. FDA appr.
   b. Brown liquid. M.W. 201.30; S.G. 0.95.
   Staining.

**19**
1. Butyraldehyde - acetaldehyde - aniline reaction product
2. —
3. A-100 — Monsanto
4. Dark red-brown, oily liquid. S.G. 1.01–1.07.
   Acc. for NR, hard rubber, synth. rubbers.

**20**
1. Homologous acroleines - aromatic bases condensation products
2. —
3. Vulkacit 576 — Bayer
4. Reddish-brown liquid. S.G. 0.99.
   Staining acc. for NR, IR, BR, SBR, NBR.

## Thioureas

**21**
1. Mixed di-alkyl thiourea
2. —
3. Pennzone L — Pennwalt
   Pennzone L — Vondelingenplaat
4. Amber-coloured liquid. S.G. 1.00.
   Acc. for CR, epichlorohydrinrubber, and for mixtures of CR and chlorobutylrubber.

**22**
1. N,N'-Dibutylthiourea
2. $(C_4H_9NH)_2CS$
3. Accelerator DBT — BASF
   DBTU — Prochim
   — — Degussa
   Pennzone B — Pennwalt
   Pennzone B — Vondelingenplaat
   Robac DBTU — Robinson
4. Off-white powder. M.W. 188.3; M.P. 65 °C.
   Acc. for mercaptan-modified CR; act. for EPDM and NR; antidegr. for NR-latex and thermoplastic SBR.
   △ Act. 35
   △ Antidegr. 109

**23**
1. 1,3-Diethylthiourea
2. $(C_2H_5NH)_2CS$
3. DETU — Prochim
   JOR 4050 — Bozzetto
   — — Degussa
   Pennzone E — Pennwalt a
   Pennzone E — Vondelingenplaat a
   RC Granulat DETU (80 % DETU, 20 % saturated hydrocarbons and special dispersion agents) — Rhein Chemie
   Robac DETU — Robinson

4. Yellow powder. M.W. 132.2; M.P. 75 °C.; S.G. 0.98.
   a. White flakes. S.G. 1.12.
   Acc. for mercaptan-modified CR (Neoprene W); antidegr. for NR, NBR, SBR, CR.
   Water soluble.
   △ Antidegr. 110

**24**

1. Dimethyl-ethyl-thiourea
2. $C_2H_5NHCSN(CH_3)_2$
3. Thiate B — Vanderbilt
4. Reddish-brown liquid. M.W. 132; S.G. 1.05.
   Acc. for Neoprene-W compounds.

**25**

1. Sym. Diphenyl-thiourea
   syn. Thiocarbanilide
2. $C_6H_5.NH.CS.NH.C_6H_5$
3. 

| | |
|---|---|
| A–L Thiocarbanilide | Monsanto |
| DPTU | Prochim |
| Eveite TC | ACNA Montecatini |
| Nocceler C | Ouchi Shinko |
| Soxinol C | Sumitomo |
| Stabilisator C (formerly Vulkacit CA) | Bayer |
| — | Degussa |
| Vulcafor TC | ICI |

4. Cream-white powder. M.W. 228; M.P. 149 °C. min.; S.G. 1.31.
   Non-staining sec. acc. for CR, EPDM.
   △ Act. 37

**26**

1. N,N'-Di-ortho-tolyl-thiourea
2. 

CH3
H
N
C=S
N
H
CH3

3. Eveite DOT — ACNA Montecatini
   Nocceler DOTU — Ouchi Shinko
4. Grayish-white powder. M.W. 256.36; M.P. 149 °C.
   Acc. for NR, CR.

**27**

1. Ethylene-thiourea
   syn. 2-Mercaptoimidazoline
2. 

```
H2C—N
 |   ||
H2C  C-SH
  \ /
   N
   |
   H
```

3. 

| | |
|---|---|
| Accelerator MI 12 | Metallgesellschaft |
| — | Degussa |
| E.T.U. | Hasselt |
| E.T.U. | Prochim |
| JOR 4022 | Bozzetto |
| JOR 4022 oleato (83 % JOR 4022, 17 % paraffin oil) | Bozzetto |
| Na-22 | Du Pont |
| Na-22D (80 % dispersion of NA-22 in oil) | Du Pont |
| Nocceler 22 | Ouchi Shinko |
| Pennac CRA | Pennwalt |
| Pennac CRA | Vondelingenplaat |
| Robac 2.2 | Robinson |
| Rodanin S-62 | Dimitrova |
| Sanceler 22 | Sanshin |
| Soxinol 22 | Sumitomo |
| Vulkacit NPV/C (coated) | Bayer |

4. White powder. M.W. 102.16; M.P. 195 °C. min.; S.G. 1.43–1.45.
   Non-staining acc. for CR; ozo for NR.
   △ Antidegr. 93

**28** 1. Tetramethylthiourea

2. 
```
    N(CH3)2
   /
  C=S
   \
    N(CH3)2
```

3. NA-101 — Du Pont
4. Light tan flakes. M.W. 132; M.P. 75–80 °C.; S.G. 1.2.
Non-staining acc. for CR.

**29** 1. Trimethylthiourea

2. $CH_3NHCSN(CH_3)_2$
3. Nocceler TMU — Ouchi Shinko
Thiate E — Vanderbilt
4. Light tan flakes. M.W. 119; M.R. 68–78 °C.; S.G. 1.23.
Acc. for Neoprene W-compounds.

**Guanidine-derivatives**

**30** 1. Diarylguanidine blend

2. R1.NH.C(:NH).NH.R2
3. Accelerator 49 — Cyanamid
4. White to pinkish-white powder. M.P. ca. 134 °C.; S.G. 1.20.
Acc. for NR; act. for MBT or MBTS in SBR.
△ Act. 33

**31** 1. N,N'-Diphenylguanidine

2. $HN{=}C(NH.C_6H_5)_2$
3. 

| | |
|---|---|
| Accicure DPG | Alkali |
| Denax | Dimitrova |
| Denax DPG | Vychodočeské |
| DPG | Cyanamid |
| DPG | Anchor |
| DPG | Monsanto |
| DPG | Rhône Poulenc |
| Eveite D | ACNA Montecatini |
| Nocceler D | Ouchi Shinko |
| Pennac DPG | Pennwalt |
| Sanceler D | Sanshin |
| Soxinol D | Sumitomo |

| | |
|---|---|
| Vulcafor DPG (surface treated) | ICI |
| Vulcafor DPG | ICI (India) |
| Vulkacit D | Bayer |

4. White crystalline non-hygroscopic powder or paste. M.W. 211.26; M.R. 144–146 °C.; S.G. 1.19.
Medium acc. for use with thiazoles and sulphenamides.
△ Act. 36

**32** 1. N,N'-Di-ortho-tolylguanidine
2. $2\text{-}CH_3.C_6H_4.NH.C(:NH)NH.C_6H_4.CH_3\text{-}2$
3.

| | |
|---|---|
| DOTG | Anchor |
| DOTG | Cyanamid |
| DOTG | Du Pont |
| DOTG | Rhône Poulenc |
| Eveite DOTG | ACNA Montecatini |
| Nocceler DT | Ouchi Shinko |
| Soxinol DT | Sumitomo |
| Vulcafor DOTG | ICI |
| Vulcafor DOTG (surface coated) | ICI |
| Vulkacit DOTG | Bayer |
| Vulkacit DOTG/C (surface coated) | Bayer |

4. White powder. M.W. 239; M.R. 167–173 °C.; S.G. 1.19.
Slow curing acc. for NR, SBR, NBR; act. for acidic and neutral acc.
△ Act. 38

**33** 1. Di-ortho-tolylguanidine salt of dicatecholborate
2. [Structural formula: two $CH_3$-substituted phenyl rings, each –NH– attached to C=N(H)(H) with + charge; counter-anion in brackets with – charge: two catechol rings bonded via O,O to B (O>B<O)]

3. Nocceler PR — Ouchi Shinko
   Permalux — Du Pont
4. Grayish-brown powder. M.W. 482.8; M.P. 165 °C.; S.G. 1.25.
   Non-staining acc. for Neoprene G-types; antidegr. for NR, SBR.
   △ Antidegr. 116

**34**
1. ortho-Tolylbiguanide
2. $H_2N–C(:NH)–NH–C(:NH)–NH–C_6H_4.CH_3$
3. Accelerator 80 — Du Pont
   Eveite 1000 — ACNA Montecatini
   Nocceler BG — Ouchi Shinko
   Vulkacit 1000 — Bayer
   Vulkacit 1000/C (surface coated) — Bayer
4. White powder. M.W. 191.24; M.P. 140 °C.; S.G. 1.2.
   Acc. for NR, IR, BR, SBR, NBR; act. for zinc dithiocarbamates, thiuram-, mercapto-, and sulphenamide-type acc.
   △ Act. 40

**35**
1. N,N',N''-Triphenylguanidine
2. $C_6H_5–NH–C(=N–C_6H_5)–NH–C_6H_5$
3. Vulcafor TPG — ICI †
4. White amorphous powder. M.W. 279; M.P. 142 °C.; S.G. 1.17.

## Xanthates

**36**
1. Sodium isopropylxanthate
2. $(H_3C)_2 > CHO–C(=S)–S–Na$
3. Aero Xanthate 343 — Cyanamid a
   Sanceler SX — Sanshin b
   Vulcafor SPX — ICI †

4. a. Light orange powder. M.W. 158; M.P. 126 °C.; S.G. 1.40.
b. Hydrate (2 $H_2O$). M.W. 194.25; M.P. 124 °C.
Water-soluble ultra-acc. for NR, NBR, SBR, CR. Also for latex.

**37**
1. Zinc dibutylxanthate
2. $(H_9C_4.O.CS.S-)_2Zn$
3. Nocceler ZBX — Ouchi Shinko
Sanceler ZBX — Sanshin
4. White powder. M.W. 363.88; M.P. 105 °C. min.; S.G. 1.24–1.40.
Ultra-acc. for NR, NBR, SBR. Also for latex. Non-staining.

**38**
1. Zinc diethylxanthate
2. $(CH_3.CH_2.O.CS.S-)_2Zn$
3. Eveite XZ — ACNA Montecatini
4. White powder. M.W. 307.

**39**
1. Zinc diisopropylxanthate
2. $[(CH_3)_2CH.O.CS.S-]_2Zn$
3. Accelerator ZIX — Anchor
Nocceler ZIX — Ouchi Shinko
Robac ZIX — Robinson
Sanceler ZX — Sanshin
Vulcafor ZIX — ICI †
4. White powder. M.W. 335.83. M.P. 145 °C. S.G. 1.54.
Fast acc. at room-temperature for NR, CR, NBR, SBR. Also for latex.

## Dithiocarbamates

**40**
1. Activated dithiocarbamates
2. —
3. Ancazate SM — Anchor
Ancazate XX — Anchor
Butyl Eight — Vanderbilt
Merac — Pennwalt
Merac — Vondelingenplaat
Robac Gamma — Robinson

Royalac 133 — Uniroyal
Setsit 5 — Vanderbilt
Setsit 9 — Vanderbilt
Setsit 51 — Vanderbilt
Trimate L (50 % active) — Vanderbilt

4. Amber-to-brown liquids. S.G. ca. 1.00.
Room-temp. acc. spec. for CR latices.

**41**
1. Bismuth dimethyldithiocarbamate
2. $[(CH_3)_2N.CS.S-]_3Bi$
3. BDMC — Hasselt
Bismate — Vanderbilt
JO 6000 — Bozzetto
Robac Bi.D.D. — Robinson
4. Yellow powder and pellets. M.W. 569.66; M.P. above 227 °C. (with decomposition); S.G. 2.02.
Acc. for NR, SBR, IIR; act. for thiazole- and sulphenamide-type acc.
△ Act. 18

**42**
1. Cadmium diethyldithiocarbamate
2. $[(C_2H_5)_2N.CS.S-]_2Cd$
3. Cadmate — Vanderbilt a
Robac C.D.C. — Robinson b
4. Cream coloured powder. M.W. 409; S.G. 1.39
a. M.R. 68–76 °C.
b. M.P. 242 °C.
Ultra-acc. for IRR. Acc. for EPDM.

**43**
1. Cadmium pentamethylenedithiocarbamate
2. $[(CH_2)_5N.CS.S-]_2Cd$
3. Robac C.P.D. — Robinson
4. Pale yellow powder. M.W. 433; M.P. 260 °C.
Del.act. acc. for NR, SBR. (cements and proofing compounds).

**44**
1. Copper dimethyldithiocarbamate
2. $[(CH_3)_2N.CS.S-]_2Cu$

3. CDMA — Hasselt
   Cumate — Vanderbilt
   Hermat Cu — Dimitrova
   JO 4015 — Bozzetto
   Nocceler TTCU — Ouchi Shinko
   Robac Cu.D.D. — Robinson
   Sanceler TTCU — Sanshin
   Soxinol MK — Sumitomo
4. Dark brown powder. M.W. 303.98; M.P. above 325 °C. (with decomposition); S.G. 1.75.
   Acc. for SBR, IIR; act. for thiazole- and sulphenamide-type acc.
   △ Act. 19

**45**
1. 2-Benzothiazyl-N,N-diethyldithiocarbamate
   syn. 2-Benzothiazyl-N,N-diethylthiocarbamylsulphide
2. (benzothiazole ring)–S–C(=S)–N($C_2H_5$)($C_2H_5$)
3. Ethylac — Pennwalt
   Ethylac — Vondelingenplaat
   Nocceler 64 — Ouchi Shinko
4. Yellow powder. M.W. 282.45; M.P. ca. 70 °C.; S.G. 1.27.
   Non-staining acc. and act. for NR, BR, SBR, NBR, IR.
   △ Act. 17

**46**
1. Dibutylammonium-dibutyldithiocarbamate
2. $(C_4H_9)_2.N.CS.S.NH_2(C_4H_9)_2$
3. Robac D.B.U.D. — Robinson
4. Brown flakes. M.W. 334.6; M.P. 45 °C.
   Acc. for NR and SBR. (proofing compounds); oxi for rubber-based adhesives. Non-staining.
   △ Antidegr. 100

**47**
1. Diethylammonium-diethyldithiocarbamate
2. $(C_2H_5)_2N.CS.S.NH_2(C_2H_5)_2$
3. Ancazate WSE — Anchor a
   Vulcafor DDCN — ICI †

4. Cream crystalline powder. M.W. 208; M.P. 80 °C.; S.G. 1.11.
   a. 30 % aquous solution. Yellow liquid. S.G. 1.028.
   Fast acc. for latex compounds.

**48**
1. Dimethylammonium-dimethyldithiocarbamate
2. $[(CH_3)_2N.CS.S-]NH_2(CH_3)_2$
3. Pennac DDD — Pennwalt
   Vondac DADD — Vondelingenplaat
4. 40 % aquous solution. Pale yellow liquid. S.G. 1.055.
   Acc. for latex.

**49**
1. N,N-Dimethylcyclohexamine salt of dibutyldithiocarbamic acid
2. —
3. RZ-100 — Monsanto
4. Yellow-orange lumps. Crystallization Point 52.5 °C. min.
   Non-staining acc. for curing at room temperature of NR- and SBR cements and latices.

**50**
1. 2,4-Dinitrophenyl-dimethyldithiocarbamate
2. $(CH_3)_2N.CS.S.C_6H_3(NO_2)_2$–2,4
3. Safex — Uniroyal
4. Yellow crystalline powder. M.W. 287; M.R. 140–145 °C.; S.G. 1.57.
   Del.act. acc.; act. for thiazole-type acc.
   △ Act. 20

**51**
1. Ferric dimethyldithiocarbamate
2. $[(CH_3)_2N.CS.S-]_3Fe$
3. Nocceler TTFE — Ouchi Shinko
   Sanceler TTFE — Sanshin
4. Black-brown powder, M.W. 416.50; M.P. 230 °C. min.
   Ultra-acc. for NBR, SBR.

**52**
1. Lead dimethyldithiocarbamate
2. $[(CH_3)_2N.CS.S-]_2Pb$
3. JO 4014 — Bozzetto
   LDMC — Hasselt
   Ledate — Vanderbilt
   Robac L.M.D. — Robinson

4. White powder. M.W. 447.65; M.P. above 310 °C.; S.G. 2.43.
Acc. for NR, BR, IR, SBR, IIR; act. for thiazole- and sulphenamide-type acc.
△ Act. 21

**53**
1. Lead pentamethylenedithiocarbamate
2. $[(CH_2)_5N.CS.S-]_2Pb$
3. Robac L.P.D. — Robinson
4. Off-white powder. M.W. 527.8; M.P. 230 °C.
Del.act. acc. for NR, SBR. (continuous vulcanization of cable-compounds).

**54**
1. Piperidinium-pentamethylenedithiocarbamate
syn. N-Pentamethyleneammonium-N-pentamethylenedithiocarbamate
2. $(CH_2)_5N.CS.S.H_2N(CH_2)_5$
3. Accelerator 552 — Du Pont
Accelerator 2P — Anchor
Nocceler PPD — Ouchi Shinko
Pentalidine — Prochim
Robac P.P.D. — Robinson
Vulkacit P — Bayer
4. Cream powder. M.W. 246; M.P. 175 °C.; S.G. 1.19.
Acc. for NR, NBR, SBR; act. for thiuram- and thiazole-type acc.; pept. for Neoprene G and KNR types.
△ Act. 22
△ Pept. 6

**55**
1. Pipecolin methyl-pentamethylenedithiocarbamate
2. $CH_3C_5H_9NC(=S)-SH-N(H)-C_5H_9CH_3$
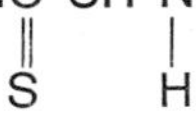
3. Nocceler P — Ouchi Shinko
4. M.W. 274.49; M.P. 120 °C.
Acc. for NR, IR, BR, SBR, NBR, IIR.

**56**
1. Potassium di-N-butyldithiocarbamate
2. $(C_4H_9)_2N.CS.S.K$
3. Butyl Kamate — Vanderbilt
4. Straw coloured aquous-solution. S.G. 1.10.
Acc. for latex.

**57** 1. Selenium diethyldithiocarbamate

2. $[(C_2H_5)_2N.CS.S-]_4Se$

3. Ethyl Selenac — Vanderbilt

Ethyl Seleram SA-66-1 (oiled) — Pennwalt

Seleniame — Prochim

Soxinol SE — Sumitomo

4. Yellow-orange powder. M.W. 672; M.R. 59–85 °C.; S.G. 1.32.

Acc. for IIR; vulc. for NR, NBR, SBR; act. for thiazole-type acc.

△ Act. 23

△ Vulc. 40

**58** 1. Selenium dimethyldithiocarbamate

2. $\left[ \begin{matrix} CH_3 \\ CH_3 \end{matrix} \!\!>\! N-\underset{\underset{S}{\parallel}}{C}-S- \right]_4 Se$

3. Methyl Selenac — Vanderbilt

4. Yellow powder and rods. M.W. 559.78; M.R. 140–172 °C.; S.G. 1.58.

Acc. and vulc. for NR, IIR, SBR, BR, IR.

△ Vulc. 41

**59** 1. Sodium dibutyldithiocarbamate

2. $(C_4H_9)_2N.CS.S.Na$

3. Ancazate WSB (48 % aquous sol.) — Anchor

Butyl Soderame (47 % aquous sol.) — Prochim

Nocceler TP — Ouchi Shinko

Robac S.B.U.D. (45 % aquous sol.) — Robinson

SBTC (40 % aquous sol.) — Bozzetto

Soxinol TP (40 % aquous sol.) — Sumitomo

Tepidone (47 % aquous sol.) — Du Pont

4. Clear brown liquid. M.W. 227; S.G. 1.075–1.09.
Ultra-acc. for latices and for reclaimed mixes; act. for thiazole-type acc.
△ Act. 24

**60** 1. Sodium diethyldithiocarbamate
2. $(C_2H_5)_2N.CS.S.Na$
3.

| | |
|---|---|
| Ethyl Soderame | Prochim a |
| Eveite L | ACNA Montecatini b |
| Nocceler SDC | Ouchi Shinko a |
| Pennac SDED (25 % aquous sol.) | Pennwalt |
| Robac S.E.D. (23 % aquous sol.) | Robinson |
| Sanceler ES (20–22 % aquous sol.) | Sanshin |
| Soxinol ESL (18–22 % aquous sol.) | Sumitomo |
| Super Accelerator 1500 | Rhône Poulenc c |
| Vondac SDED (25 % aquous sol.) | Vondelingenplaat |
| Vulcafor SDC | ICI † |

4. a. White crystalline powder. M.W. 171.26; S.G. 1.3.
b. White crystalline powder. M.W. 207.29; M.P. 90–92 °C.; S.G. 1.30.
c. Hydrate. M.W. 225; S.G. 1.30.
Ultra-acc. for NR- and SBR latices; act. for guanidine-type acc. Non-staining.
△ Act. 25

**61** 1. Sodium dimethyldithiocarbamate
2. $(CH_3)_2N.CS.S.Na$
3.

| | |
|---|---|
| Eveite K | ACNA Montecatini |
| Methyl Soderame | Prochim |
| Nocceler S | Ouchi Shinko |
| Sanceler S | Sanshin |
| Soxinol MSL | Sumitomo |

4. 40–42 % aquous-solution, light orange coloured. M.W. 143.20; S.G. 1.17–1.19.
   Ultra-acc. for NR- and SBR latices in combination with other acc.

**62**
1. Tellurium diethyldithiocarbamate
2. $[(C_2H_5)_2N.CS.S-]_4Te$
3. Soxinol TE — Sumitomo
   TDEC — Hasselt
   Tellurac — Vanderbilt
   Tellurame — Prochim
4. Orange-yellow powder and rods. M.W. 720.69; M.R. 108–118 °C.; S.G. 1.44.
   Acc. for NR, SBR, NBR, IIR, EPDM; act. for thiazole- and thiuram-type acc.
   △ Act. 26

**63**
1. Zinc dithiocarbamate (activated)
2. —
3. Ancazate Q — Anchor
4. Viscous brown liquid. S.G. 1.10.
   Room-temperature acc. for NR, SBR.

**64**
1. Zinc dibenzyldithiocarbamate
2. $[(C_6H_5CH_2)_2N.CS.S-]_2Zn$
3. Arazate — Uniroyal a
   Robac Z.B.E.D. — Robinson b
   ZBEC — Hasselt b
4. Creamy-white powder. M.W. 610.2; S.G. 1.41.
   a. M.R. 160–175 °C.
   b. M.P. 182 °C. min.
   Ultra-acc. for IIR, SBR, NR. Also for latex. Act. for thiazole- and sulphenamide-type acc.
   △ Act. 27

**65**
1. Zinc dibutyldithiocarbamate
2. $[(C_4H_9)_2N.CS.S-]_2Zn$

3. Aceto ZDBD — Aceto
   Ancazate BU — Anchor
   Butazate — Naugatuck SpA
   Butazate — Uniroyal
   Butazate 50D — Uniroyal
   (50 % slurry for use in latex)
   Butazin — Ticino
   Butyl Zimate — Vanderbilt
   Butyl Ziram — Pennwalt
   Butyl Ziram — Pennwalt
   (50 % aquous solution)
   Butyl Zirame — Prochim
   Eptac 4 — Du Pont
   Eveite Butil Z — ACNA Montecatini
   JO 4013 — Bozzetto
   Nocceler BZ — Ouchi Shinko
   Robac Z.B.U.D. — Robinson
   Sanceler BZ — Sanshin
   Soxinol BZ — Sumitomo
   Ultra Accelerator Di 13 — Metallgesellschaft
   Vondac ZBUD — Vondelingenplaat
   Vulcafor ZNBC — ICI †
   ZDBC — Hasselt
4. Cream powder. M.W. 474.14; M.R. 95–108 °C.; S.G. 1.21.
   Non-staining ultra-acc. for NR, BR, SBR, NBR and their latices, acc. for EPDM; antidegr. for unvulcanized rubber and for non-staining grades of IIR; act. for thiazole- and other acid-type acc.
   △ Act. 28
   △ Antidegr. 103

**66**
1. Zinc dibutyldithiocarbamate - dibutylamine complex
2. $[(C_4H_9)_2N.CS.S-]_2Zn\ .\ (C_4H_9)_2NH$
3. Robac Z.B.U.D.X. — Robinson
4. Golden brown liquid. M.W. 603.4.
   Non-staining acc. for solutions and for cements.

**67**
1. Zinc diethyldithiocarbamate
2. $[(C_2H_5)_2N.CS.S-]_2Zn$

3. | Trade name | Manufacturer |
|---|---|
| Accicure ZDC | Alkali |
| Aceto ZDED | Aceto |
| Ancazate ET | Anchor |
| Etazin | Ticino |
| Ethasan | Monsanto |
| Ethazate | Naugatuck SpA |
| Ethazate | Uniroyal |
| Ethazate 50 D (50 % slurry) | Uniroyal |
| Ethyl Zimate | Vanderbilt |
| Ethyl Ziram (oiled) | Pennwalt |
| Ethyl Ziram (50 % aquous dispersion) | Pennwalt |
| Ethyl Zirame | Prochim |
| Eveite Z | ACNA Montecatini |
| Hermat ZDK | Dimitrova |
| Nocceler EZ | Ouchi Shinko |
| Robac ZDC | Robinson |
| Sanceler EZ | Sanshin |
| Soxinol EZ | Sumitomo |
| Superaccelerator 1505 | Rhône Poulenc |
| Superaccelerator 1505 N | Rhône Poulenc a |
| Ultra Accelerator Di 7 | Metallgesellschaft |
| Vondac ZDC | Vondelingenplaat |
| Vulcafor ZDC | ICI |
| Vulcafor ZDC | ICI (India) |
| Vulkacit LDA | Bayer |
| ZDEC | Hasselt |

4. White powder. M.W. 361.91; M.P. 175 °C. min.; S.G. 1.49.
   Ultra-acc. for NR, SBR, NBR, IIR, and their latices; act. for thiazole-type acc. Non-staining.
   a. Special for use in latex.
   △ Act. 29

**68** 1. Zinc dimethyldithiocarbamate
2. $[(CH_3)_2N.CS.S-]_2Zn$

3. Aceto ZDMD — Aceto
   Ancazate ME — Anchor
   Cyazate M — Cyanamid
   Eptac 1 — Du Pont
   Eveite Metil Z — ACNA Montecatini
   Hermat ZDM — Dimitrova
   JO 4012 — Bozzetto
   JO 4012 Oleato (83 % JO 4012, 17 % paraffin oil) — Bozzetto
   Metazin — Ticino
   Methasan — Monsanto
   Methazate — Naugatuck SpA
   Methazate — Uniroyal
   Methyl Zimate — Vanderbilt
   Methyl Ziram (oiled and extruded) — Pennwalt
   Methyl Ziram (50 % dispersion) — Pennwalt
   Methyl Zirame — Prochim
   Nocceler PZ — Ouchi Shinko
   Rhenocure ZMC — Rhein Chemie
   Robac ZMD — Robinson
   Sanceler PZ — Sanshin
   Soxinol PZ — Sumitomo
   Superaccelerator 1605 — Rhône Poulenc
   Ultra accelerator Di 4 — Metallgesellschaft
   Vondac ZMD — Vondelingenplaat
   Vulkacit L — Bayer
   ZDMC — Hasselt
4. White yellowish powder. M.W. 305.82; M.P. ca. 250 °C.; S.G. 1.66.
   Non-staining ultra-acc. for NR, SBR, IIR and their latices; act. for thiazole- and sulphenamide-type acc.
   △ Act. 30

**69**
1. Zinc dimethyl-pentamethylenedithiocarbamate
2. $[C_5H_8(CH_3)_2N.CS.S-]_2Zn$

3. Robac Z.L. — Robinson
4. Yellow-white powder. M.W. 442; M.P. 64 °C.
   Non-staining acc. for latex.

**70**
1. Zinc ethyl-phenyldithiocarbamate
2. $[C_6H_5(C_2H_5)N.CS.S-]_2Zn$
3. 

| Trade name | Manufacturer |
|---|---|
| Ancazate EPH | Anchor |
| Eveite P | ACNA Montecatini |
| Hermat FEDK | Dimitrova |
| Nocceler PX | Ouchi Shinko |
| Sanceler PX | Sanshin |
| Soxinol PX | Sumitomo |
| Superaccelerator 1105 | Rhône Poulenc |
| Vondac ZEPD | Vondelingenplaat |
| Vulcafor ZEP | ICI |
| Vulkacit P extra N | Bayer |

4. White-yellow powder. M.W. 458.02; M.P. 205 °C.; S.G. 1.50.
   Ultra-acc. for NR, IIR, SBR and their latices.

**71**
1. Zinc N-pentamethylenedithiocarbamate
2. $(C_5H_{10}N.CS.S-)_2Zn$
3. 

| Trade name | Manufacturer |
|---|---|
| Robac Z.P.D. | Robinson |
| Vulkacit ZP | Bayer |
| ZPMC | Hasselt |

4. Off-white powder. M.W. 385.9; M.P. 225 °C.; S.G. 1.60.
   Non-staining ultra-acc. for latex in combination with other zinc-dithiocarbamates; act. for thiazole- and sulpenamide-type acc.
   △ Act. 31

**72**
1. Zinc pentamethylenedithiocarbamate - piperidine complex
2. $(C_5H_{10}N.CS.S-)_2Zn . C_5H_{10}NH$
3. Robac Z.P.D.X. — Robinson
4. White powder. M.W. 471.1; M.P. 140 °C.
   Also available as a 50 % paste.
   Acc. and act. for M.B.T.
   △ Act. 32

**Thiuramsulphides**

**73**
1. Cyclic thiuram-type products
2. —
3. Conac T — Du Pont a
   Sanceler CT — Sanshin b
4. a. Yellow powder. M.P. 96.1 °C.; S.G. 1.26.
   Acc. for SBR.
   b. Yellow powder. M.P. 115 °C. min.
   Del.act. acc. for SBR, NBR, IIR, IR, EPT.

**74**
1. Dimethyl-diphenyl-thiuram disulphide
2. $C_6H_5(CH_3)N.CS.S.S.CS.N(CH_3)C_6H_5$
3. DDTS — Bozzetto
   Vulkacit J — Bayer
4. White crystalline powder. M.W. 364; M.P. 175 °C. min.; S.G. 1.33.
   Non-staining del.act. acc. for NR, SBR, IR, BR, NBR. Sec. acc. for TMTS.

**75**
1. Dipentamethylene-thiuram disulphide
2. $(C_5H_{10}N.CS.S)_2$
3. Robac P.T.D. — Robinson a
   Robac P.T.D. 86 — Robinson b
   (containing extra sulphur)
4. Cream coloured powder. M.W. 320.6;
   a. M.P. 120 °C.
   b. M.P. 110 °C.
   Non-staining acc. and vulc. for latex (gloves) and for IIR (pharmaceutical closures).
   △ Vulc. 32

**76**
1. Dipentamethylene-thiuram tetrasulphide
2. $(CH_2)_5N.CS.S_4.CS.N(CH_2)_5$
3. Accelerator 4P — Anchor
   DPTT — Hasselt
   Nocceler TRA — Ouchi Shinko
   Robac P.25 — Robinson
   Soxinol TRA — Sumitomo

4. Light yellow powder or rods. M.W. 384.69; M.P. 115 °C.; S.G. 1.50.
   Ultra-acc. and vulc. for CSM, IIR, EPDM, NBR, SBR, IR, CR, NR; act. for thiazole- and sulphenamide-type acc.
   △ Act. 7
   △ Vulc. 34

**77**
1. Dipentamethylene-thiuram hexasulphide
2. 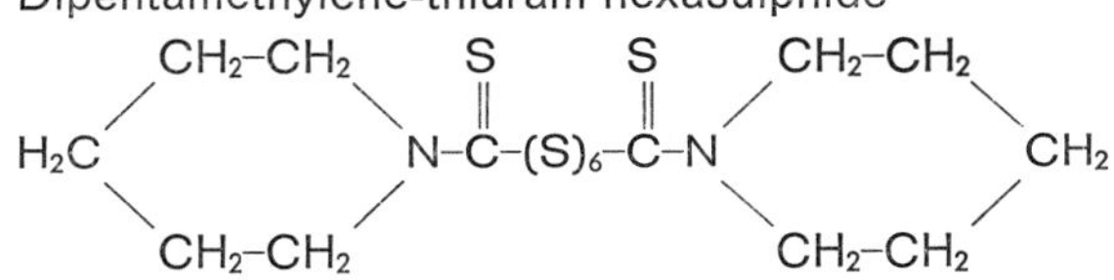

3. DPTT — Akron
   Sulfads — Vanderbilt
   Tetrone A — Du Pont
4. Light gray powder. M.W. 448; M.P. 110 °C.; S.G. 1.53.
   Non-staining acc. for NR, SBR, NBR, CR, IR, IIR, EPDM, CSM. Also for latex. Vulc. for IR.
   △ Vulc. 33

**78**
1. Dipentamethylene-thiuram monosulphide
2. $(C_5H_{10}N.CS)_2S$
3. Robac P.T.M. — Robinson
4. Yellow powder. M.W. 288.5; M.P. 110 °C.
   Del.act. acc. for chemically-blown cellular NR and SBR products; acc. for NBR in combination with MBTS.

**79**
1. Tetrabutyl-thiuram disulphide
2. $[(C_4H_9)_2N.CS.S-]_2$
3. Robac T.B.U.T. — Robinson
   Soxinol TBT — Sumitomo
   TBTS — Bozzetto
4. Brown liquid. M.W. 408.7; solidifies at ca. 20 °C.; S.G. 1.1. Insoluble in water.
   Non-staining acc. in combination with P.T.D. or T.M.T. for NR, SBR, NBR in sulphurless compounds; ret. for CR; vulc. for NR, SBR, NBR.
   △ Ret. 10
   △ Vulc. 35

**80**
1. Tetrabutyl-thiuram monosulphide
2. $(C_4H_9)_2N.CS.S.CS.N(C_4H_9)_2$
3. Pentex — Uniroyal a
   Pentex Flour — Uniroyal b
   (Pentex 12.5 %, clay 87.5 %)
4. a. Brown liquid. M.W. 376; S.G. 0.99.
   Del.action acc. for NR.
   b. Buff coloured powder. S.G. 2.16.
   Acc. for spongerubber.

**81**
1. Tetraethyl-thiuram disulphide
2. $(C_2H_5)_2N.CS.S.S.CS.N(C_2H_5)_2$
3.

| | |
|---|---|
| Accicure TET | Alkali |
| Aceto TETD | Aceto |
| Ancazide ET | Anchor |
| Ethyl Thiram | Pennwalt |
| Ethyl Thiurad | Monsanto |
| Ethyl Tuads | Vanderbilt a |
| Ethyl Tuex | Naugatuck SpA |
| Ethyl Tuex | Uniroyal |
| Etiurac | Ticino |
| Eveite T | ACNA Montecatini |
| Hermat TET | Dimitrova |
| Nocceler TET | Ouchi Shinko |
| Robac TET | Robinson |
| Sanceler TET | Sanshin |
| Soxinol TET | Sumitomo |
| Super Accelerator 481 | Rhône Poulenc |
| TETD | Hasselt |
| T.E.T.D. | Prochim |
| TETS | Bozzetto |
| Thiuram E | Du Pont |
| Vondac TET | Vondelingenplaat |
| Vulcafor TET | ICI |
| Vulcafor TET | ICI (India) |

4. Grayish white powder and pellets. M.W. 296.54; M.P. 71–73 °C.; S.G. 1.26.
   a. Powder and rods.

Non-staining acc. for NR, SBR, NBR, BR, IIR, IR, EPDM; act. for thiazole-, guanidine- and aldehyde-type acc.; vulc. for sulphurless compounds; stabilizer for Neoprene GN.

△ Act. 8

△ Antidegr. 96

△ Vulc. 36

**82**

1. Tetramethyl-thiuram monosulphide
2. $(CH_3)_2N.CS.S.CS.N(CH_3)_2$
3. | | |
   |---|---|
   | Aceto TMTM | Aceto |
   | Ancazide IS | Anchor |
   | Cyuram MS | Cyanamid |
   | Eveite MST | ACNA Montecatini |
   | Monex | Naugatuck SpA |
   | Monex | Uniroyal |
   | Mono Thiurad | Monsanto |
   | Nocceler TS | Ouchi Shinko |
   | Pennac MS | Pennwalt |
   | Robac TMS | Robinson |
   | Sanceler TS | Sanshin |
   | Soxinol TS | Sumitomo |
   | Superaccelerator 500 | Rhône Poulenc |
   | Thionex | Du Pont |
   | TMTM | Hasselt |
   | Unads | Vanderbilt |
   | Vulcafor MS | ICI |
   | Vulkacit Thiuram MS | Bayer |
   | Vulkacit Thiuram MS/C (surface coated) | Bayer |
4. Yellow powder, pellets or rods. M.W. 208.37; M.R. 103–114 °C.; S.G. 1.37.

   Non-staining ultra-acc. for NR, NBR, IIR; act. for mercapto- and sulphenamide-type acc.

   △ Act. 10

**83**

1. Tetramethylthiuram disulphide
2. $(CH_3)_2N.CS.S.S.CS.N(CH_3)_2$

| | | |
|---|---|---|
| 3. | Accicure TMT | Alkali |
| | Aceto TMTD | Aceto |
| | Ancazide ME | Anchor |
| | Cyuram DS | Cyanamid |
| | Eveite 4MT | ACNA Montecatini |
| | Hermat TMT | Dimitrova |
| | Methyl Thiram (oiled and extruded) | Pennwalt |
| | Methyl Tuads | Vanderbilt a |
| | Metiurac | Ticino |
| | Nocceler TT | Ouchi Shinko |
| | RC Granulat TMTD (80 % TMTD, 20 % saturated hydrocarbons and dispersing agents) | Rhein Chemie |
| | Robac TMT | Robinson |
| | Sanceler TT | Sanshin |
| | Soxinol TT | Sumitomo |
| | Superaccelerator 501 | Rhône Poulenc |
| | Thiurad | Monsanto |
| | Thiuram 16 | Metallgesellschaft |
| | Thiuram M | Du Pont |
| | TMTD | Hasselt |
| | TMTD | Akron |
| | TMTD | Prochim |
| | TMTS | Bozzetto |
| | TMTS oleato (83 % TMTS, 17 % paraffin oil) | Bozzetto |
| | Tuex | Naugatuck SpA |
| | Tuex | Uniroyal |
| | Vondac TMT | Vondelingenplaat |
| | Vulcafor TMT | ICI |
| | Vulcafor TMT | ICI (India) |
| | Vulkacit Thiuram | Bayer |
| | Vulkacit Thiuram C (surface coated) | Bayer |
| | Vulkacit Thiuram GR (granules) | Bayer |

4. White to yellow powder. M.W. 240.44; M.P. 146–148 °C.; S.G. 1.3–1.4.
   a. Powder and rods.
   Non-staining ultra-acc. and vulc.; act. for thiazole- and sulphenamide-type acc.
   △ Act. 9
   △ Vulc. 37

**84** 1. Tetraethyl-thiuram disulphide / tetramethyl-thiuram disulphide blend
2. —
3. Methyl Ethyl Tuads — Vanderbilt a
   Pennac TM — Pennwalt b
4. a. 50:50 mixture. White to cream rods. M.P. 62 °C. min.; S.G. 1.32.
   b. Light buff powder. M.W. 257.5; S.G. 1.24.
   Acc. and vulc. for sulphur-less or low-sulphur stocks of NR, SBR, EPDM.
   △ Vulc. 38

## Heterocyclic compounds

**85** 1. 4,4'-Dithiodimorpholine
   syn. Dimorpholinyl disulphide
   Morpholine disulphide
2. $O(CH_2CH_2)_2N{-}S{-}S{-}N(CH_2CH_2)_2O$
3. Deovulc M — Grandel
   Sulfasan R — Monsanto
   Vanax A — Vanderbilt
   Vondac DTDM — Vondelingenplaat
   Vulnoc R — Ouchi Shinko
4. White to grey powder. M.W. 236.36; M.R. 115–127 °C.; S.G. 1.35. FDA appr.
   Vulc. and acc. for NR, SBR, NBR, IIR, EPDM.
   △ Vulc. 44

**86** 1. 2-Mercaptobenzimidazole

2. [structure: benzimidazole-2-thiol (C–SH, N, NH) ⇄ benzimidazoline-2-thione (NH, C=S, NH)]

3.

| | |
|---|---|
| Antigene MB | Sumitomo |
| Antioxidant MB | Dimitrova |
| Antioxidant MB | Bayer |
| Antivecchiante MB | ACNA Montecatini |
| MBI | Prochim |
| Nocrac MB | Ouchi Shinko |
| Permanax 21 | Rhône Poulenc |
| Vondantox MBI | Vondelingenplaat |

4. Yellow-white powder. M.W. 150; M.P. 290 °C. (with decomposition); S.G. 1.42.
Acc. for NR; non-staining oxi, heat, inh for NR, SBR, NBR; pept. for CR.
△ Antidegr. 94
△ Pept. 3

**87** 1. Carbon disulphide - 1,1'-methylene-dipiperidine reaction product

2. —

3. R.2 Crystals — Monsanto

4. Gray-white pellets. M.P. 55 °C. min.; S.G. 1.08–1.14.
Ultra-acc. for latex and fast-curing cements. FDA appr.

**88** 1. Thiohydropyrimidine

2. —

3. Thiate A — Vanderbilt

4. White crystalline powder. M.P. 250 °C. min.; S.G. 1.12.
Acc. for CR.

**Thiazoles**

**89** 1. Benzothiazyl-2-diethylsulphenamide

2. [structure: benzothiazole (N, S) –S–$N(C_2H_5)_2$]

3. Vulkacit AZ — Bayer
4. Dark-brown oily liquid, hygroscopic. M.W. 238; S.G. 1.18.
   Del. action acc. for NR, IR, BR, SBR, NBR.

**90**
1. 1,3-Bis-(2-benzothiazyl mercaptomethyl) urea
2. N N
   $-S-CH_2-NH-C-NH-CH_2-S-$
   S O S
3. El-Sixty — Monsanto
4. Buff powder. M.W. 418.63; M.P. 220 °C.; S.G. 1.35–1.41.
   Acc. for compounds containing large amounts of retarding fillers, requires ZnO and fatty acid; sec. acc. for ultra- and thiazole-type acc. in EPDM. Non-staining.

**91**
1. N-Cyclohexylamine salt of 2-mercaptobenzothiazole
2. N $CH_2-CH_2$
   $-S-H.H_2N-CH$ $CH_2$
   S $CH_2-CH_2$
3. Nocceler M-60 — Ouchi Shinko
   Sanceler HM — Sanshin
4. Yellowish powder. M.W. 266.43; M.P. 150 °C. min.
   Ultra-acc. for NR, SBR, NBR. Also for latex.

**92**
1. 2,2'-Dibenzothiazyl disulphide
   syn. Di-(2-benzothiazyl) disulphide
2. 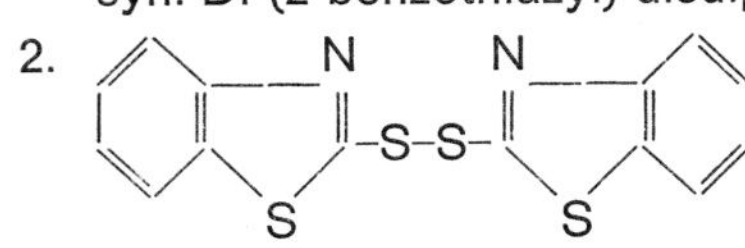

3. Accicure MBTS — Alkali
   Altax — Vanderbilt
   Ancatax — Anchor
   Bowax AC/MBTS — Bozzetto
   (MBTS dispersed in Bowax C)
   Eveite DM — ACNA Montecatini
   MBTS — Akron

| | |
|---|---|
| MBTS | Cyanamid |
| MBTS | Du Pont |
| MBTS | Naugatuck SpA |
| MBTS | Uniroyal |
| Mercasulf MBTS | Bozzetto |
| Nocceler DM | Ouchi Shinko |
| Pennac MBTS | Pennwalt |
| Pneumax DM | Dimitrova |
| Pneumax F | Dimitrova a |
| Rapid Accelerator 201 | Rhône Poulenc |
| Sanceler DM | Sanshin |
| Soxinol DM | Sumitomo |
| Thiofide MBTS | Monsanto |
| Vulcafor MBTS | ICI |
| Vulcafor MBTS | ICI (India) |
| Vulkacit DM | Bayer |

4. Yellowish powder. M.W. 332.50; M.P. 170–175 °C.; S.G. 1.50.
   a. Mixture with basic acc.
   Non-staining semi-ultra-acc. for NR, NBR, IIR, SBR. Also for latex. Ret. for CR.
   △ Ret. 3

**93** 1. N,N-Dicyclohexyl-2-benzothiazyl sulphenamide

2.

$CH_2$–$CH_2$
CH $CH_2$
N
–S–N $CH_2$–$CH_2$
S
$CH_2$–$CH_2$
CH $CH_2$
$CH_2$–$CH_2$

3.

| | |
|---|---|
| Nocceler DZ | Ouchi Shinko |
| Vulkacit DZ | Bayer |

4. Coarse-grained brown powder. M.W. 346.58; M.P. 90 °C. min.; S.G. 1.20.
   Del.action acc. for NR, IR, BR, SBR, NBR.

**94** 1. 2-(2,6-Dimethyl-4-morpholinyl mercapto) benzothiazole

2\. [structural formula: benzothiazole–S–N (2,6-dimethylmorpholine), with $CH_3$ groups]

3\. Santocure 26 — Monsanto

4\. Brown flakes. M.W. 280.42; M.P. 84 °C. min.; S.G. 1.23–1.29.
Non-staining del.action acc. for natural and synthetic rubbers.

**95** 1. 2-(2,4-Dinitrophenyl) mercaptobenzothiazole
syn. 2-(2,4-Dinitrophenyl) thiobenzothiazole

2\. [structural formula: benzothiazole–S–phenyl with two $NO_2$ groups]

3\. Eveite 303 — ACNA Montecatini
Nocceler DBM — Ouchi Shinko
Ureka Base — Monsanto

4\. Yellow powder. M.W. 333.35; M.P. 155 °C.; S.G. 1.61.
Staining acc. for NR, IR, BR, SBR, NBR.

**96** 1. 2-Mercaptobenzothiazole

2\. [structural formula: benzothiazole with C.SH]

3\.

| | |
|---|---|
| Accicure MBT | Alkali |
| Ancap | Anchor |
| Captax | Vanderbilt |
| Eveite M | ACNA Montecatini |
| MBT | Akron |
| MBT | Cyanamid |
| MBT XXX (very pure grade) | Cyanamid |
| MBT | Du Pont |
| MBT | Naugatuck SpA |
| MBT | Uniroyal |
| Nocceler M | Ouchi Shinko |

| | |
|---|---|
| Pennac MBT | Pennwalt |
| Pneumax MBT | Dimitrova |
| Rapid Accelerator 200 | Rhône Poulenc |
| Rotax | Vanderbilt |
| Sanceler M | Sanshin |
| Soxinol M | Sumitomo |
| Thiotax (MBT) | Monsanto |
| Vulcafor MBT | ICI |
| Vulcafor MBT | ICI (India) |
| Vulkacit Merkapto | Bayer |

4. Light yellow powder. M.W. 167.25; M.R. 164–175 °C; S.G. 1.50. Non-staining semi-ultra acc. for NR, SBR, NBR, IIR. Also for latex. Ret. for CR; pept. for NR.
   △ Pept. 4
   △ Ret. 6

**97**
1. 2-Mercaptothiazoline
2. 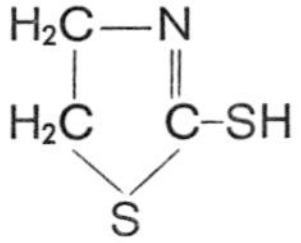 $H_2C$—N<br>$H_2C$ C–SH<br>S
3. Accelerator 2MT — Anchor
4. White powder. M.W. 119; M.P. 104–105 °C.

**98**
1. 4-Morpholinyl-2-benzothiazyl disulphide
   syn. 2-(4-Morpholino-dithiobenzothiazole)
2. 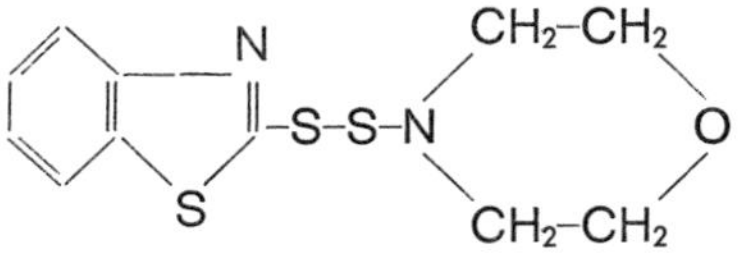 N, S –S–S–N ($CH_2$–$CH_2$, $CH_2$–$CH_2$) O
3. Morfax — Vanderbilt
   Nocceler MDB — Ouchi Shinko
4. Cream powder. M.W. 284.35; M.R. 123–135 °C.; S.G. 1.51. Acc. for NR, SBR, BR, IIR, NBR, CR, EPDM.

**99**
1. Sodium 2-mercaptobenzothiazole
2. 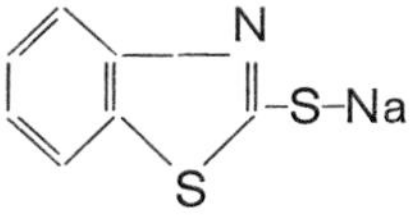 N, S –S–Na

3. Sanceler M-NA — Sanshin
4. Light-yellow crystalline powder. M.W. 189.24; M.P. 280 °C.
   Semi-ultra-acc. for NR latex.

**100** 1. Zinc-2-mercaptobenzothiazole

2.

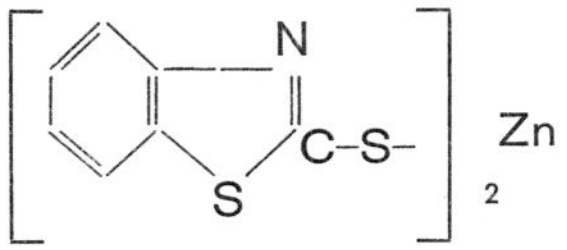

$$\left[ \text{C}_6\text{H}_4\text{(N=C(-S-)S)} \right]_2 \text{Zn}$$

3.

| | |
|---|---|
| Accicure ZMBT | Alkali |
| Bantex | Monsanto |
| Eveite MZ | ACNA Montecatini |
| Hermat Zn MBT | Dimitrova |
| Nocceler MZ | Ouchi Shinko |
| OXAF | Naugatuck SpA |
| OXAF | Uniroyal |
| Pennac ZT | Pennwalt |
| Pennac ZT-"W" (with 10 % hydrocarbon) | Pennwalt |
| Rapid Accelerator 205 | Rhône Poulenc |
| Sanceler MZ | Sanshin |
| Soxinol MZ | Sumitomo |
| Vulcafor ZMBT | ICI |
| Vulcafor ZMBT | ICI (India) |
| Vulkacit ZM | Bayer |
| Zenite (with hydrocarbon) | Du Pont |
| Zenite Special | Du Pont |
| Zetax | Vanderbilt |
| Zinc Ancap | Anchor |
| ZMBT | Cyanamid |
| ZMBT waxed | Cyanamid |
| ZMBT wettable | Cyanamid |

4. Light cream powder. M.W. 397.9; M.P. 300 °C. (with decomposition); S.G. 1.72.
   Acc. for NR-, SBR-, NBR-latices; non-staining oxi for latices.
   △ Antidegr. 99

## Sulphenamides

**101**

1. 2-(4-Morpholinyl-mercapto)-benzothiazole
   syn. N-Oxydiethylene-2-benzothiazole sulphenamide
   Benzothiazyl-2-sulphene morpholide
2. 
3. 

| | |
|---|---|
| Amax | Vanderbilt |
| Delac MOR | Uniroyal |
| NOBS Special | Cyanamid |
| Nocceler MSA | Ouchi Shinko |
| OBTS | Akron |
| Pennac MBS | Pennwalt |
| Sanceler NOB | Sanshin |
| Santocure MOR | Monsanto |
| Santocure MOR-90 (Santocure MOR/Thiofide 90 : 10) | Monsanto |
| Soxinol NBS | Sumitomo |
| Vulcafor BSM | ICI |
| Vulcafor BSM | ICI (India) |
| Vulkacit MOZ | Bayer |

4. Tan flakes. M.W. 252.3; M.R. 70–90 °C.; S.G. 1.37.
   Non-staining del.action acc. for NR, IR, BR, SBR, NBR. FDA appr.

**102**

1. N-tert. Butyl-benzothiazyl sulphenamide
2. 

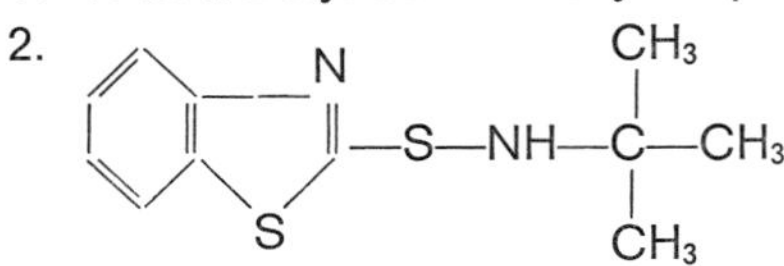


3. 

| | |
|---|---|
| BBTS | Akron |
| Conac NS | Du Pont |
| Delac NS | Uniroyal |
| Pennac TBBS | Pennwalt |
| Sanceler NS | Sanshin |

| | |
|---|---|
| Santocure NS | Monsanto |
| Santocure NS 50 (mixture with a pre-vulc. inhibitor) | Monsanto |
| Soxinol NS | Sumitomo |
| Vulcafor BSB | ICI (India) |
| Vulkacit NZ | Bayer |

4. Light-tan or buff powder or pellets. M.W. 238.37; M.P. 104 °C. min.; S.G. 1.26–1.32.
   Non-staining acc. for NR, SBR, IR, BR. FDA appr.

**103**

1. N-Cyclohexyl-2-benzothiazyl sulphenamide
2. Structural formula: benzothiazole ring (N, S) –S–NH–CH< cyclohexane ring ($CH_2$–$CH_2$, $CH_2$, $CH_2$–$CH_2$)
3.

| | |
|---|---|
| Accicure HBS | Alkali |
| CBTS | Akron |
| Conac S | Du Pont |
| Cydac | Cyanamid |
| Delac S | Naugatuck SpA |
| Delac S | Uniroyal |
| Durax | Vanderbilt |
| Eveite MS | ACNA Montecatini |
| Nocceler CZ | Ouchi Shinko |
| Pennac CBS | Pennwalt |
| Rhodifax 16 | Rhône Poulenc |
| Sanceler CM | Sanshin |
| Santocure | Monsanto |
| Soxinol CZ | Sumitomo |
| Sulfenax CB/30 | Dimitrova |
| Vulcafor HBS | ICI |
| Vulcafor HBS | ICI (India) |
| Vulkacit CZ | Bayer |

4. Greenish-tan flakes. M.W. 264.41; M.R. 90–108 °C.; S.G. 1.27–1.30.
   Del.action acc. for NR, SBR. Non-staining.

**104** 1. N,N-Diisopropyl-2-benzothiazyl sulphenamide
2. $C_6H_4\langle_{S}^{N}\rangle C{-}S{-}N\langle_{CH(CH_3)_2}^{CH(CH_3)_2}$
3.

| | |
|---|---|
| DIBS | Cyanamid |
| DIBS | Anchor |
| Dipac | Pennwalt |
| Nocceler PSA | Ouchi Shinko |
| Sanceler DIB | Sanshin |

4. Light tan flakes. M.W. 266.42; M.R. 55–60 °C.; S.G. 1.21. Del.action acc. for NR, SBR, IR, BR.

**Miscellaneous, mixtures and undisclosed compositions**

**105** 1. Amine-salt of a dialkyl dithiophosphoric acid
2. —
3. Rhenocure AT — Rhein Chemie
4. White crystalline powder. M.R. 98–104 °C.; S.G. 1.04. Acc. for EPDM. Non-staining.

**106** 1. Bis-(2-ethylamino-4-diethylamino-triazine-6-yl) disulphide.
2. Two triazine rings, each bearing $NH{-}C_2H_5$ and $N(C_2H_5)_2$ groups, joined by S—S: $(C_2H_5)_2N$, N, N, N, $NH{-}C_2H_5$ … S—S … $NH{-}C_2H_5$, N, N, N, $N(C_2H_5)_2$
3. Accelerator KA 9032 — Bayer
4. Cream-coloured powder. M.W. 428; M.P. ca. 105 °C.; S.G. 1.126. Del.action acc. for NR, IR, SBR, NBR, BR.

**107** 1. N-Cyclohexyl-2-benzothiazyl sulphenamide / diphenylguanidine blend (70 : 30)
2. —
3. Pneumax CB/N — Dimitrova
4. Acc. for natural and synthetic rubbers.

**108** 1. 2,2'-Dibenzothiazyl disulphide / basic accelerators blend
2. —

3. Eveite F — ACNA Montecatini
   Vulkacit F — Bayer
   Vulkacit F/C — Bayer
   (surface coated)
4. Yellowish powder. M.P. 150 °C. min.; S.G. 1.31.
   Acc. for NR, IR, BR, SBR, NBR.

**109**
1. 2,2'-Dibenzothiazyl disulphide / diphenylguanidine blend
2. —
3. Accicure F — Alkali
   Vulcafor F — ICI
   (surface coated)
   Vulcafor F — ICI (India)
4. Cream-white powder. S.G. 1.46.

**110**
1. 2,2'-Dibenzothiazyl disulphide / hexamethylenetetramine blend
2. —
3. Nocceler Mix 3 — Ouchi Shinko
   Soxinol G3 — Sumitomo
4. Yellowish-white powder. M.P. 160 °C. min.
   Acc. for NR, SBR, NBR.

**111**
1. 2,2'-Dibenzothiazyl disulphide / 2-morpholinyl-thio-benzothiazyl blend (10 : 90)
2. —
3. Pennac MBS 90 — Pennwalt
4. Tan chips. S.G. 1.36.

**112**
1. 2,2'-Dibenzothiazyl disulphide / N-oxydiethylene-2-benzothiazyl sulphenamide blend.
2. —
3. Amax No. 1 — Vanderbilt
   NOBS No. 1 — Cyanamid
   (90 : 10)
4. Tan flakes. M.R. 70–90 °C.; S.G. 1.40.
   Del.action acc. for use with high pH furnace blacks in NR and SBR.

**113** 1. 2,2'-Dibenzothiazyl disulphide / diphenylguanidine / hexamethylenetetramine blend.
2. —
3. Accicure FN — Alkali
Nocceler F — Ouchi Shinko
Pneumax H/N — Dimitrova
Sanceler F — Sanshin
Soxinol F — Sumitomo
Vulcafor FN — ICI
4. Yellow powder. M.P. 145 °C.; S.G. 1.3–1.4.
Non-staining acc. for NR, SBR, IR, BR, NBR.

**114** 1. 2,2'-Dibenzothiazyl disulphide / thiuram disulphide blend.
2. —
3. Vulcafor DAU — ICI †
4. Cream powder. S.G. 1.47.

**115** 1. Dibutyl-xanthogen disulphide
2. $C_4H_9{-}O{-}\underset{\underset{S}{\|}}{C}{-}S{-}S{-}\underset{\underset{S}{\|}}{C}{-}O{-}C_4H_9$
3. C.P.B. — Uniroyal
4. Amber-coloured liquid. M.W. 298; S.G. 1.15.
Non-staining acc. for low-temperature vulcanization of NR, SBR, NBR, IIR, CR.

**116** 1. 2,4-Dinitrophenyl benzothiazyl sulphide / diphenylguanidine blend.
2. —
3. Pneumax U — Dimitrova
4. Yellow powder. M.P. 120 °C.
Acc. for natural and synthetic rubbers.

**117** 1. Dithiocarbamate / tetraethyl-thiuram disulphide blend (2 : 1)
2. —
3. Royalac 235 — Uniroyal
4. Black powder. S.G. 1.46.
Acc. for EPDM.

**118** 1. Glycol-dimercaptoacetate
2. $HSCH_2COO(CH_2)_2OCOCH_2SH$
3. Robac G.D.M.A. — Robinson
4. Pale-yellow liquid. M.W. 210.3.
Acc. for chlorobutylrubber. Non-staining.

**119** 1. Mercapto-derivatives / guanidines blends
2. —
3. Nocceler U — Ouchi Shinko
4. M.P. 125 °C.
Acc. for NR, SBR, NBR.

**120** 1. 2,2'-Dibenzothiazyl disulphide / hexamethylenetetramine / 2-mercaptobenzothiazole blend
2. —
3. Nocceler Mix 2 — Ouchi Shinko
Soxinol G2 — Sumitomo
4. Light-yellow powder. M.P. 115 °C. min.
Acc. for NR, SBR, NBR.

**121** 1. 2-Mercaptobenzothiazole / dithiocarbamate blend
2. —
3. Accicure DHC — Alkali
Vulcafor DHC — ICI
Vulcafor DHC — ICI (India)
4. Pale-yellow powder. S.G. 1.47.

**122** 1. 2-Mercaptobenzothiazole / ferric diethyldithiocarbamate blend
2. —
3. Nocceler EP-40 — Ouchi Shinko
4. M.P. 140 °C.
Acc. for EPDM.

**123** 1. 2-Mercaptobenzothiazole / ferric dimethyldithiocarbamate blend
2. —
3. Nocceler EP 10 — Ouchi Shinko
4. M.P. 190 °C.
Acc. for EPDM.

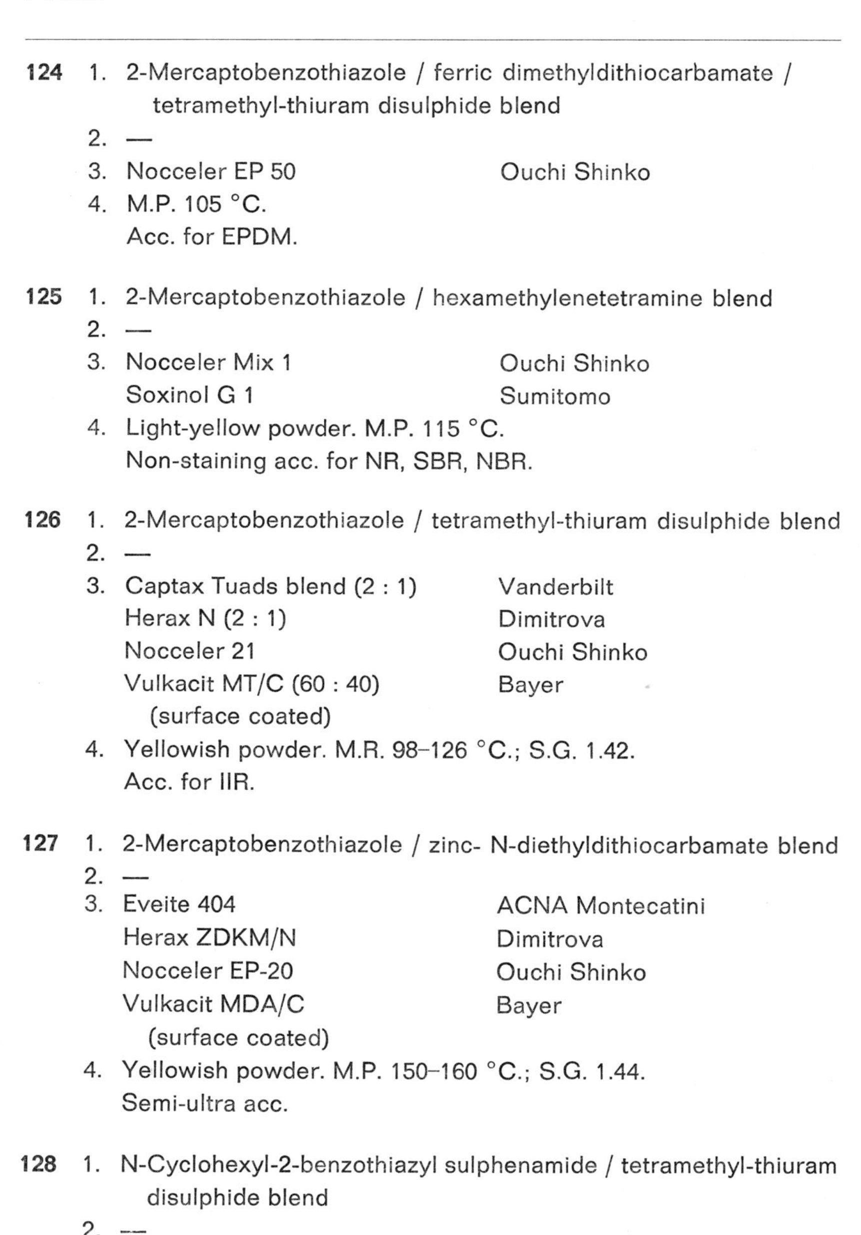

**124** 1. 2-Mercaptobenzothiazole / ferric dimethyldithiocarbamate / tetramethyl-thiuram disulphide blend

2. —

3. Nocceler EP 50 — Ouchi Shinko

4. M.P. 105 °C.
Acc. for EPDM.

**125** 1. 2-Mercaptobenzothiazole / hexamethylenetetramine blend

2. —

3. Nocceler Mix 1 — Ouchi Shinko
Soxinol G 1 — Sumitomo

4. Light-yellow powder. M.P. 115 °C.
Non-staining acc. for NR, SBR, NBR.

**126** 1. 2-Mercaptobenzothiazole / tetramethyl-thiuram disulphide blend

2. —

3. Captax Tuads blend (2 : 1) — Vanderbilt
Herax N (2 : 1) — Dimitrova
Nocceler 21 — Ouchi Shinko
Vulkacit MT/C (60 : 40) (surface coated) — Bayer

4. Yellowish powder. M.R. 98–126 °C.; S.G. 1.42.
Acc. for IIR.

**127** 1. 2-Mercaptobenzothiazole / zinc- N-diethyldithiocarbamate blend

2. —

3. Eveite 404 — ACNA Montecatini
Herax ZDKM/N — Dimitrova
Nocceler EP-20 — Ouchi Shinko
Vulkacit MDA/C (surface coated) — Bayer

4. Yellowish powder. M.P. 150–160 °C.; S.G. 1.44.
Semi-ultra acc.

**128** 1. N-Cyclohexyl-2-benzothiazyl sulphenamide / tetramethyl-thiuram disulphide blend

2. —

3. Herax CB/X/N — Dimitrova

4. Acc. for natural and synthetic rubbers.

**129** 1. 2,2'-Dibenzothiazyl disulphide / bis-(2-ethylamino-4-diethylamino-triazine-6-yl) sulphide / tetramethyl-thiuram disulphide blend (33 : 33 : 34)
2. —
3. Accelerator KA 9029 Bayer
4. White-yellow powder. M.P. ca. 90 °C.; S.G. 1.4.
Acc. for NR, IR, SBR, NBR, BR.

**130** 1. 2,2'-Dibenzothiazyl disulphide / bis-(2-ethylamino-4-diethylamino-triazine-6-yl) disulphide / tetramethyl-thiuram disulphide blend (46 : 46 : 8)
2. —
3. Accelerator KA 9030 Bayer
4. Yellow-white powder. M.P. ca. 100 °C.; S.G. 1.11.
Acc. for NR, IR, SBR, NBR, BR.

**131** 1. Tetramethyl-thiuram disulphide / zinc dimethyldithiocarbamate blend (50 : 50)
2. —
3. Eptac 2 Du Pont
Nocceler EP-30 Ouchi Shinko
Royalac 134 Uniroyal
4. Off-white powder. S.G. 1.52.
Non-staining acc. for EPDM, also for latex.

**132** 1. Bis-(2-ethylamino-4-diethylamino-triazine-6-yl) disulphide / tetramethyl-thiuram monosulphide blend (92 : 8)
2. —
3. Accelerator KA 9031 Bayer
4. Yellow-white powder. M.P. ca. 100 °C.; S.G. 1.26.
Acc. for NR, IR, SBR, NBR, BR.

**133** 1. Tetramethyl-thiuram monosulphide / zinc mercaptobenzothiazole blend (3 : 97)
2. —
3. Zenite A Du Pont
Zenite AM Du Pont
(micronized Zenite A)
4. Cream-coloured powder. S.G. 1.53.
Non-staining acc. for NR, SBR, NBR, IIR.

**134** 1. Thiazole / acid salt of diphenylguanidine blend
2. —
3. Vulcafor DAT — ICI a †
Vulcafor DAW — ICI b †
4. a. Greyish powder. M.P. 127–130 °C.
b. Yellow powder. S.G. 1.40.

**135** 1. Thiazole / guanidine blend
2. —
3. Blendac F — Anchor a
Blendac FM — Anchor †
Blendac V — Anchor b
Ureka White — Monsanto c
4. Pale-cream powder. Non-staining.
a. S.G. 1.43, acc. for NR, SBR.
b. S.G. 1.46, acc. for NR, SBR, NBR.
c. S.G. 1.39.

**136** 1. Thiazole / guanidine / hexamine blend
2. —
3. Blendac VN — Anchor
4. Cream powder. S.G. 1.47.
Non-staining fast del.action acc.

**137** 1. Thiuram / thiazole blends
2. —
3. Accicure BT — Alkali a
Blendac R — Anchor b
Vulcafor BT — ICI (India) a
4. a. Cream-coloured powder. S.G. 1.44.
b. Cream-coloured powder. S.G. 1.37.
Acc. for NR, IIR.

**138** 1. Zinc o,o-di-n-butylphosphorodithioate
2. —
3. Vocol — Monsanto
Vocol S — Monsanto
(62 % active ingredient,
38 % silicious carrier)

4. Clear yellow-green liquid. Flash Point 195 °C.; S.G. 1.25.
Non-staining acc. for EPDM.

**139** 1. Zinc salt of dialkyldithiophosphoric acid.
2. —
3. Rhenocure TP — Rhein Chemie a
Rhenocure TP/S — Rhein Chemie b
(Mixture of Rhenocure TP and $SiO_2$, 67 : 33)
4. a. Greenish-yellow liquid. S.G. 1.25.
Acc. for EPDM.
b. White powder. S.G. 1.42.
Acc. for EPDM.

**140** 1. Silicic acid / zinc salt of a thiophosphoricester blend
2. —
3. Deovulc S — Grandel
4. White powder. S.G. 1.4.
Acc. for EPDM.

**141** 1. Composition undisclosed
2. —
3. Diak Super 6 — Du Pont
Eveite 202 — ACNA Montecatini
Harmon — Ticino
Kenmix — Kenrich a
Pennac NB liquid — Pennwalt
Pennac NB powder — Pennwalt
(74 % active ingredient on inert carrier)
Pennac NB liquid — Vondelingenplaat
Pennac powder — Vondelingenplaat
Rapid Accelerator 465 — Rhône Poulenc
Rhodifax 7 — Rhône Poulenc
Rhodifax 10 — Rhône Poulenc
Rhodifax 14 — Rhône Poulenc
Robac 70 — Robinson
Robac Alpha — Robinson
Robac C.S. — Robinson

| | |
|---|---|
| Royalac 136 | Uniroyal |
| Soxinol RL-13 | Sumitomo |
| TA 11 | Du Pont |
| Triox | Ticino |

4. a. △ Act. 44
   △ Antidegr. 125
   △ Vulc. 54

# Activators

## INORGANIC

**1**

1. Lead oxides
2. —
3. Mix Lpb 80 — Bozzetto a
   Mix Pb 80 — Bozzetto b
   Polyminium — Polychimie b
   Polytharge — Polychimie a
   grades A, B, D
   RC Granulat PbO — Rhein Chemie a
   RC Granulat $Pb_3O_4$ — Rhein Chemie b
   — — Anchor c
4. a. PbO dispersion. Acc. for CR; act. for NR, NBR, SBR, IIR; vulc. for CR, CSM.
   b. $Pb_3O_4$ dispersion. Acc. for CR; act. for IIR.
   c. PbO powder (litharge). Vulc. for CR.
   △ Acc. 1
   △ Vulc. 1

**2**

1. Magnesium oxide
2. MgO
3. Kenmag — Kenrich a
   Maglite — Merck b
   grades D, L, K, M, Y
   RC Granulat MgO — Rhein Chemie c
   (80 % MgO, 20 % saturated hydrocarbons and dispersion agents)
   Scorchguard — Anchor d
   grades C3, O, W
   Scorchguard O — Newalls e
   Struktol — Schill & Seilacher f
   grades WB 900, WB 902 (coated)

4. M.W. 40.32.
   a. S.G. 2.02. Act. for CR.
   b. White powders. S.G. 3.3–3.5.
      Act. for SBR and fluoro-elastomers; vulc. for CR, CSM; antidegr. for CR, CSM, chlorobutyl, fluoro-elastomers, SBR.
   c. S.G. 2.06.
   d. Grade C3, powder, heavy calcined MgO.
      Grade O, putty, light calcined MgO.
      Grade W, powder, light calcined MgO.
   e. Dispersion. S.G. 2.08.
   f. Act. for CR and CSM mixtures.
   △ Antidegr. 1
   △ Vulc. 2

**3**

1. Magnesium oxide / zinc oxide blends
2. —
3. Struktol WB 890 — Schill & Seilacher
4. White powder, 45 % ZnO, 36 % MgO, 19 % dispersion agents. Act. for CR.

**4**

1. Siliciumdioxide / surface active agents blends
2. —
3. Aktivator 1987 — Rhein Chemie a
   Aktivator 2009 — Rhein Chemie a
   Aktivator 2642 — Rhein Chemie a
   Aktivator R — Degussa b
4. a. White powder. S.G. 1.5–1.75.
      Act. for vulcanizers and for blowing agents.
   b. Mixture of Ultrasil and hexanetriol.
      White powder.
   Acceleration-act. for SBR.

**5**

1. basic Zinccarbonate
2. $ZnCO_3.2ZnO.3H_2O$
3. Zinkoxid Transparent — Bayer
   — — Durham

4. White powder. S.G. 3.3–3.5.
   Act. for sulphur- and peroxide vulcanization of NR, IR, BR, SBR, NBR, IIR, CR, EPDM, CSM; vulc. for CR.
   △ Vulc. 7

**6**

1. Zinc oxide
2. ZnO
3. | | |
   |---|---|
   | Activox | Durham a |
   | Decelox | Durham a |
   | Durox 25 | Durham a |
   | Flocculent | Durham a |
   | Loled | Durham a |
   | Microx | Durham a |
   | Mix Zn 60 | Bozzetto b |
   | Mix Zn 60 paste | Bozzetto c |
   | Noled | Durham a |
   | RC Granulat ZnO (80 % ZnO, 20 % saturated hydrocarbons and dispersion agents) | Rhein Chemie d |
   | RC Zinkoxid 64 (90 % ZnO, 10 % dispersion agents) | Rhein Chemie d |
   | S.R.Q. | Durham a |
   | Struktol Neopast | Schill & Seilacher e |
   | Struktol WB 700 | Schill & Seilacher f |
   | Zinkoxid Aktiv | Bayer a |
   | — | ACNA Montecatini a |
   | — | Anchor g |
4. a. White powder. M.W. 81. S.G. 5.4.
   b. 60 % micronized ZnO in microcrystalline waxes and hydrocarbon resins. White flakes. M.P. 50 °C., S.G. 1.7–1.8.
   c. White paste. M.P. 64–68 °C., S.G. 1.85–1.9.
   d. Yellow-white powder. S.G. 4.0.
   e. Grey paste. Act. for CR- and CSM mixtures. S.G. 2.1.
   f. White powder. Act. for CR- and CSM mixtures.
   g. Act. for CR.

## ORGANIC

### Thiuramsulphides

**7**

1. Dipentamethylene-thiuram tetrasulphide
2. $(CH_2)_5N.CS.S_4.CS.N(CH_2)_5$
3. Accelerator 4P — Anchor
   DPTT — Hasselt
   Nocceler TRA — Ouchi Shinko
   Robac P.25 — Robinson
   Soxinol TRA — Sumitomo
4. Light yellow powder or rods. M.W. 384.69; M.P. 115 °C.; S.G. 1.50.
   Act. for thiazole- and sulphenamide-type acc.; ultra-acc. and vulc. for CSM, IIR, EPDM, NBR, SBR, IR, CR, NR.
   △ Acc. 76
   △ Vulc. 34

**8**

1. Tetraethyl-thiuram disulphide
2. $(C_2H_5)_2.N.CS.S.S.CS.N(C_2H_5)_2$
3. Accicure TET — Alkali
   Aceto TETD — Aceto
   Ancazide ET — Anchor
   Ethyl Thiram, extruded — Pennwalt
   Ethyl Thiurad — Monsanto
   Ethyl Tuads — Vanderbilt a
   Ethyl Tuex — Naugatuck SpA
   Ethyl Tuex — Uniroyal
   Etiurac — Ticino
   Eveite T — ACNA Montecatini
   Hermat TET — Dimitrova
   Nocceler TET — Ouchi Shinko
   Robac TET — Robinson
   Sanceler TET — Sanshin
   Soxinol TET — Sumitomo
   Superaccelerator 481 — Rhône Poulenc
   TETD — Prochim
   TETD — Hasselt
   TETS — Bozzetto

| | |
|---|---|
| Thiuram E | Du Pont |
| Vondac TET | Vondelingenplaat |
| Vulcafor TET | ICI |
| Vulcafor TET | ICI (India) |

4. Grayish white powders and pellets. M.W. 296.54; M.P. 71–73 °C.; S.G. 1.26.

   a. Powder and rods.

   Act. for thiazole-, guanidine- and aldehyde-type acc.; acc. for NR, SBR, NBR, BR, IIR, IR, EPDM; vulc. for sulphurless compounds; stabilizer for Neoprene GN. Non-staining.

   △ Acc. 81

   △ Antidegr. 96

   △ Vulc. 36

**9**

1. Tetramethyl-thiuram disulphide
2. $(CH_3)_2N.CS.S.S.CS.N(CH_3)_2$
3. 

| | |
|---|---|
| Accicure TMT | Alkali |
| Aceto TMTD | Aceto |
| Ancazide ME | Anchor |
| Cyuram DS | Cyanamid |
| Eveite 4MT | ACNA Montecatini |
| Hermat TMT | Dimitrova |
| Methyl Thiram oiled | Pennwalt |
| Methyl Thiram extruded | Pennwalt |
| Methyl Tuads | Vanderbilt a |
| Metiurac | Ticino |
| Nocceler TT | Ouchi Shinko |
| RC Granulat TMTD (80 % TMTD, 20 % saturated hydrocarbons and dispersion agents) | Rhein Chemie |
| Robac TMT | Robinson |
| Sanceler TT | Sanshin |
| Soxinol TT | Sumitomo |
| Superaccelerator 501 | Rhône Poulenc |
| Thiurad | Monsanto |
| Thiuram 16 | Metallgesellschaft |
| Thiuram M | Du Pont |

| | |
|---|---|
| TMTD | Akron |
| TMTD | Hasselt |
| TMTD | Prochim |
| TMTS | Bozzetto |
| TMTS oleato (83 % TMTS, 17 % paraffin oil) | Bozzetto |
| Tuex | Naugatuck SpA |
| Tuex | Uniroyal |
| Vondac TMT | Vondelingenplaat |
| Vulcafor TMT | ICI |
| Vulcafor TMT | ICI (India) |
| Vulkacit Thiuram | Bayer |
| Vulkacit Thiuram C (surface coated) | Bayer |
| Vulkacit Thiuram GR (granules) | Bayer |

4. White to yellow powder. M.W. 240.44; M.P. 146–148 °C.; S.G. 1.3–1.4.

   a. Powder and rods.

   Act. for thiazole- and sulphenamide-type acc.; non-staining ultra-acc. and vulc.

   △ Acc. 83

   △ Vulc. 37

**10**

1. Tetramethyl-thiuram monosulphide
2. $(CH_3)_2N.CS.S.CS.N.(CH_3)_2$
3. 

| | |
|---|---|
| Aceto TMTM | Aceto |
| Ancazide IS | Anchor |
| Cyuram MS | Cyanamid |
| Eveite MST | ACNA Montecatini |
| Monex | Naugatuck SpA |
| Monex | Uniroyal |
| Mono Thiurad | Monsanto |
| Nocceler TS | Ouchi Shinko |
| Pennac MS | Pennwalt |
| Robac TMS | Robinson |
| Sanceler TS | Sanshin |
| Soxinol TS | Sumitomo |

| | |
|---|---|
| Superaccelerator 500 | Rhône Poulenc |
| Thionex | Du Pont |
| TMTM | Hasselt |
| Unads | Vanderbilt |
| Vulcafor MS | ICI |
| Vulkacit Thiuram MS | Bayer |
| Vulkacit Thiuram MS/C (surface coated) | Bayer |

4. Yellow powder, pellets or rods. M.W. 208.37; M.R. 103–114 °C.; S.G. 1.37.
   Act. for mercapto- and sulphenamide-type acc.; non-staining ultra-acc. for NR, NBR, IIR.
   △ Acc. 82

**Amines**

**11**

1. Butyraldehyde - aniline condensation product
2. —
3. Accelerator 21 — Anchor

| | |
|---|---|
| Accelerator 21 | Anchor |
| Antox Special | Du Pont |
| Beutene | Uniroyal |
| Butanyl 1 | Ticino |
| Nocceler 8 | Ouchi Shinko |
| Rapid Accelerator 300A | Rhône Poulenc |
| Vulcafor BA | ICI |

4. Orange-red oily liquid. S.G. 0.94–1.02.
   Act. for thiazoles, thiurams and guanidines; semi-ultra acc. for Neoprene W and for hydrocarbon rubber; antidegr. for CR.
   △ Acc. 14
   △ Antidegr. 41

**12**

1. Cyclohexyl ethyl amine
2. $\overbrace{(CH_2)_5CH}$.NH.$C_2H_5$
3. Vulkacit HX — Bayer
4. Pale yellow liquid. M.W. 127; B.P. 165 °C.; S.G. 0.85.
   Sec. acc. and act
   △ Acc. 7

**13** 1. Dibenzylamine / monobenzylamine blend
2. —
3. D.B.A. — Uniroyal
4. Yellow liquid. S.G. 1.03.
Act. for dibutyl-xanthogen disulphide acc.

**14** 1. Heptaldehyde - aniline condensation product
2. —
3. Heptene Base — Uniroyal
4. Dark brown liquid. S.G. 0.92.
Act. for thiazole- and thiuram-type acc.; acc. for NR.
△ Acc. 17

**15** 1. Hexamethylenetetramine (with additives)
2. $(CH_2)_6N_4$
3.

| Trade name | Manufacturer |
|---|---|
| Aceto HMT | Aceto |
| Eveite UR | ACNA Montecatini |
| Herax UTS | Dimitrova |
| Hexa K | Degussa |
| Hexalit grades S, 01, 02 | Chemko |
| Nocceler H | Ouchi Shinko |
| RC Granulat Hexa (80 % hexa and 20 % saturated hydrocarbons and dispersion agents) | Rhein Chemie |
| Rhenocure Hexa | Rhein Chemie † |
| Sanceler H | Sanshin |
| Soxinol H | Sumitomo |
| Vulkacit H 30 | Bayer |

4. Hygroscopic white powder. M.W. 140.19; sublimates at 263 °C.; S.G. 1.3.
Act. for mercapto-, sulphenamide-, thiuram- and zinc dithiocarbamate-type acc.; acc. for NR, SBR, NBR.
△ Acc. 2

**16** 1. Triethanolamine
2. $N(CH_2CH_2OH)_3$

3. Ankoltet — Anchor
   TEA — Shell
   grades TEA 85 %,
   TEA Commercial, TEA 98 %
4. Water-white liquid. M.W. 149.19; B.P. 286 °C.; S.G. 1.12.
   Act. for accelerators.

## Dithiocarbamates

**17**

1. 2-Benzothiazyl-N,N-diethyldithiocarbamate
   syn. 2-Benzothiazyl-N,N-diethylthiocarbamyl sulphide
2. (benzothiazol-2-yl)–S–C(=S)–N$(C_2H_5)_2$
3. Ethylac — Pennwalt
   Ethylac — Vondelingenplaat
   Nocceler 64 — Ouchi Shinko
4. Yellow powder. M.W. 282.45; M.P. ca. 70 °C.; S.G. 1.27.
   Non-staining acc. and act. for NR, BR, SBR, NBR, IR.
   △ Acc. 45

**18**

1. Bismuth dimethyldithiocarbamate
2. $[(CH_3)_2N.CS.S-]_3Bi$
3. BDMC — Hasselt
   Bismate — Vanderbilt
   JO 6000 — Bozzetto
   Robac Bi.D.D. — Robinson
4. Yellow powder and pellets. M.W. 569.66; M.P. above 227 °C. (with decomposition); S.G. 2.02.
   Act. for thiazole- and sulphenamide-type acc.; acc. for NR, SBR, IIR.
   △ Acc. 41

**19**

1. Copper dimethyldithiocarbamate
2. $[(CH_3)_2N.CS.S-]_2Cu$
3. CDMA — Hasselt
   Cumate — Vanderbilt
   Hermat Cu — Dimitrova

| | | |
|---|---|---|
| | JO 4015 | Bozzetto |
| | Nocceler TTCU | Ouchi Shinko |
| | Robac Cu.D.D. | Robinson |
| | Sanceler TTCU | Sanshin |
| | Soxinol MK | Sumitomo |

4. Dark brown powder. M.W. 303.98; M.P. above 325 °C. (with decomposition); S.G. 1.75.
Act. for thiazole- and sulphenamide-type acc.; acc. for SBR, IIR.
△ Acc. 44

**20**
1. 2,4-Dinitrophenyl-dimethyldithiocarbamate
2. $(CH_3)_2N.CS.S.C_6H_3(NO_2)_2$–2,4
3. Safex — Uniroyal
4. Yellow crystalline powder. M.W. 287; M.R. 140–145 °C.; S.G. 1.57.
Act. for thiazole-type acc.; del. action acc.
△ Acc. 50

**21**
1. Lead dimethyldithiocarbamate
2. $[(CH_3)_2N.CS.S-]_2Pb$
3. JO 4014 — Bozzetto
LDMC — Hasselt
Ledate — Vanderbilt
Robac L.M.D. — Robinson
4. White powder. M.W. 447.65; M.P. above 310 °C.; S.G. 2.43.
Act. for thiazole- and sulphenamide-type acc.; acc. for NR, BR, IR, SBR, IIR.
△ Acc. 52

**22**
1. Piperidinium-pentamethylenedithiocarbamate
syn. N-Pentamethyleneammonium-N-pentamethylenedithio-carbamate
2. $(CH_2)_5N.CS.S.H_2N(CH_2)_5$
3. Accelerator 552 — Du Pont
Accelerator 2P — Anchor
Nocceler PPD — Ouchi Shinko
Pentalidine — Prochim
Robac P.P.D. — Robinson
Vulkacit P — Bayer

4. Cream powder. M.W. 246; M.P. 175 °C.; S.G. 1.19.
   Act. for thiuram- and thiazole-type acc.; acc. for NR, NBR, SBR; pept. for Neoprene G- and KNR types.
   △ Acc. 54
   △ Pept. 6

**23**
1. Selenium diethyldithiocarbamate
2. $[(C_2H_5)_2N.CS.S-]_4Se$
3. Ethyl Selenac — Vanderbilt
   Ethyl Seleram SA-66-1 (oiled) — Pennwalt
   Seleniame — Prochim
   Soxinol SE — Sumitomo
4. Yellow-orange powder. M.W. 672; M.R. 59–85 °C.; S.G. 1.32.
   Act. for thiazole-type acc.; acc. for IIR, vulc. for NR, NBR, SBR.
   △ Acc. 57
   △ Vulc. 40

**24**
1. Sodium dibutyldithiocarbamate
2. $(C_4H_9)_2N.CS.S.Na$
3. Ancazate WSB (48 % aquous solution) — Anchor
   Butyl Soderame (47 % aquous solution) — Prochim
   Nocceler TP — Ouchi Shinko
   Robac SBUD (45 % aquous solution) — Robinson
   SBTC (40% aquous solution) — Bozzetto
   Soxinol TP (40% aquous solution) — Sumitomo
   Tepidone (47 % aquous solution) — Du Pont
4. Clear brown liquid. M.W. 227; S.G. 1.075–1.09.
   Act. for thiazole-type acc.; ultra-acc. for latices and for reclaimed mixes.
   △ Acc. 59

**25**

1. Sodium diethyldithiocarbamate
2. $(C_2H_5)_2N.CS.S.Na$
3. | | |
   |---|---|
   | Ethyl Soderame | Prochim a |
   | Eveite L | ACNA Montecatini b |
   | Nocceler SDC | Ouchi Shinko a |
   | Pennac SDED (25 % aquous solution) | Pennwalt |
   | Robac S.E.D. (23 % aquous solution) | Robinson |
   | Sanceler ES (20–22 % aquous solution) | Sanshin |
   | Soxinol ESL (18–22 % aquous solution) | Sumitomo |
   | Super Accelerator 1500 | Rhône Poulenc c |
   | Vondac SDED (25 % aquous solution) | Vondelingenplaat |
   | Vulcafor SDC | ICI † |
4. a. White crystalline powder. M.W. 171.26; S.G. 1.3.
   b. White crystalline powder. M.W. 207.29; M.P. 90–92 °C.; S.G. 1.30.
   c. Hydrate. M.W. 225; S.G. 1.30.
   Act. for guanidine-type acc.; ultra-acc. for NR- and SBR latices. Non-staining.
   △ Acc. 60

**26**

1. Tellurium diethyldithiocarbamate
2. $[(C_2H_5)_2N.CS.S-]_4Te$
3. | | |
   |---|---|
   | Soxinol TE | Sumitomo |
   | TDEC | Hasselt |
   | Tellurac | Vanderbilt |
   | Tellurame | Prochim |
4. Orange-yellow powder and rods. M.W. 720.69; M.R. 108–118 °C.; S.G. 1.44.
   Act. for thiazole- and thiuram-type acc.; acc. for NR, SBR, NBR, IIR, EPDM.
   △ Acc. 62

**27**

1. Zinc dibenzyldithiocarbamate
2. $[(C_6H_5CH_2)_2N.CS.S-]_2Zn$

3. Arazate — Uniroyal a
   Robac Z.B.E.D. — Robinson b
   ZBEC — Hasselt b
4. Creamy-white powder. M.W. 610.2; S.G. 1.41.
   a. M.R. 160–175 °C.
   b. M.P. 182 °C. min.
   Act. for thiazole- and sulphenamide-type acc.; ultra-acc. for IIR, SBR, NR, also for latex.
   △ Acc. 64

**28**
1. Zinc dibutyldithiocarbamate
2. $[(C_4H_9)_2N.CS.S-]_2Zn$
3. Aceto ZDBD — Aceto
   Ancazate BU — Anchor
   Butazate — Naugatuck SpA
   Butazate — Uniroyal
   Butazate 50D — Uniroyal
   (50 % slurry for use in latex)
   Butazin — Ticino
   Butyl Zimate — Vanderbilt
   Butyl Ziram — Pennwalt
   Butyl Ziram — Pennwalt
   (50 % aquous solution)
   Butyl Zirame — Prochim
   Eptac 4 — Du Pont
   Eveite Butil Z — ACNA Montecatini
   JO 4013 — Bozzetto
   Nocceler BZ — Ouchi Shinko
   Robac Z.B.U.D. — Robinson
   Sanceler BZ — Sanshin
   Soxinol BZ — Sumitomo
   Ultra Accelerator Di 13 — Metallgesellschaft
   Vondac ZBUD — Vondelingenplaat
   Vulcafor ZNBC — ICI †
   ZDBC — Hasselt
4. Cream powder. M.W. 474.14; M.R. 95–108 °C.; S.G. 1.21.
   Act. for thiazole- and other acid-type acc.; non-staining ultra-acc. for NR, BR, SBR, NBR and their latices, acc. for EPDM;

antidegr. for unvulcanized rubber, and for non-staining grades of IIR.

△ Acc. 65

△ Antidegr. 103

**29**

1. Zinc diethyldithiocarbamate
2. $[(C_2H_5)_2N.CS.S-]_2Zn$
3. 

| | |
|---|---|
| Accicure ZDC | Alkali |
| Aceto ZDED | Aceto |
| Ancazate ET | Anchor |
| Etazin | Ticino |
| Ethasan | Monsanto |
| Ethazate | Naugatuck SpA |
| Ethazate | Uniroyal |
| Ethazate 50D (50 % dispersion) | Uniroyal |
| Ethyl Zimate | Vanderbilt |
| Ethyl Ziram (oiled) | Pennwalt |
| Ethyl Ziram (50 % dispersion) | Pennwalt |
| Ethyl Zirame | Prochim |
| Eveite Z | ACNA Montecatini |
| Hermat ZDK | Dimitrova |
| Nocceler EZ | Ouchi Shinko |
| Robac ZDC | Robinson |
| Sanceler EZ | Sanshin |
| Soxinol EZ | Sumitomo |
| Superaccelerator 1505 | Rhône Poulenc |
| Superaccelerator 1505(N) | Rhône Poulenc a |
| Ultra Accelerator Di 7 | Metallgesellschaft |
| Vondac ZDC | Vondelingenplaat |
| Vulcafor ZDC | ICI |
| Vulcafor ZDC | ICI (India) |
| Vulkacit LDA | Bayer |
| ZDEC | Hasselt |

4. White powder. M.W. 361.91; M.P. 175 °C. min.; S.G. 1.49.
   Act. for thiazole-type acc.; non-staining ultra-acc. for NR, SBR, NBR, IIR and their latices (a. special for use in latex).

   △ Acc. 67

**30** Zinc dimethyldithiocarbamate

2. $[(CH_3)_2N.CS.S-]_2Zn$

3.

| Trade name | Manufacturer |
|---|---|
| Aceto ZDMD | Aceto |
| Ancazate ME | Anchor |
| Cyzate M | Cyanamid |
| Eptac 1 | Du Pont |
| Eveite Metil Z | ACNA Montecatini |
| Hermat ZDM | Dimitrova |
| JO 4012 | Bozzetto |
| JO 4012 oleato (83 % JO 4012, 17 % paraffin oil) | Bozzetto |
| Metazin | Ticino |
| Methasan | Monsanto |
| Methazate | Naugatuck SpA |
| Methazate | Uniroyal |
| Methyl Zimate | Vanderbilt |
| Methyl Ziram grades extruded, oiled, 50 % dispersion | Pennwalt |
| Methyl Zirame | Prochim |
| Nocceler PZ | Ouchi Shinko |
| Rhenocure ZMC | Rhein Chemie |
| Robac ZMD | Robinson |
| Sanceler PZ | Sanshin |
| Soxinol PZ | Sumitomo |
| Superaccelerator 1605 | Rhône Poulenc |
| Ultra accelerator Di 4 | Metallgesellschaft |
| Vondac ZMD | Vondelingenplaat |
| Vulkacit L | Bayer |
| ZDMC | Hasselt |

4. White-yellowisch powder. M.W. 305.82; M.P. ca. 250 °C.; S.G. 1.66.

   Act. for thiazole- and sulphenamide-type acc.; non-staining ultra-acc. for NR, SBR, IIR and their latices.

   △ Acc. 68

**31** 1. Zinc N-pentamethylenedithiocarbamate

2. $(C_5H_{10}N.CS.S-)_2Zn$

3. Robac Z.P.D. — Robinson
   Vulkacit ZP — Bayer
   ZPMC — Hasselt
4. Off-white powder. M.W. 385.9; M.P. 225 °C.; S.G. 1.60.
   Act. for thiazole- and sulphenamide-type acc.; non-staining ultra-acc. for latex in combination with other zinc-dithio-carbamates.
   △ Acc. 71

**32**

1. Zinc pentamethylenedithiocarbamate - piperidine complex
2. $(C_5H_{10}N.CS.S-)_2Zn\ .\ C_5H_{10}NH$
3. Robac Z.P.D.X. — Robinson
4. White powder. M.W. 471.1; M.P. 140 °C. Also available as a 50 % paste.
   Acc. and act. for M.B.T.
   △ Acc. 72

## Miscellaneous, mixtures and undisclosed compositions

**33**

1. Diarylguanidine blend
2. $R_1.NH.C(:NH).NH.R_2$
3. Accelerator 49 — Cyanamid
4. White to pinkish-white powder. M.P. ca. 134 °C.; S.G. 1.20.
   Act. for MBT or MBTS in SBR; acc. for NR.
   △ Acc. 30

**34**

1. Dibutylammoniumoleate
2. $$\begin{matrix} H_3C{-}CH_2{-}CH_2{-}CH_2 \searrow \\ H_3C{-}CH_2{-}CH_2{-}CH_2 \nearrow \end{matrix} \overset{\oplus}{N}H_2 \quad \overset{\ominus}{O}{-}\underset{\underset{O}{\|}}{C}{-}(C_{17}H_{33})$$
3. Activator 1102 — Anchor
   Barak — Du Pont
   DOB — Organo Synthèse
4. Dark-amber liquid. M.W. 409; Flash Point 102 °C.; S.G. 0.88.
   Act. for thiazole-, thiuram- and sulphenamide-type acc.; act.-ret. for thiuram-type acc.
   △ Ret. 4

**35** 1. N,N'-Dibutylthiourea
2. $(C_4H_9NH)_2CS$
3. Accelerator DBT — BASF
DBTU — Prochim
— — Degussa
Pennzone B — Pennwalt
Pennzone B — Vondelingenplaat
Robac DBTU — Robinson
4. Off-white powder. M.W. 188.3; M.P. 65 °C.
Act. for EPDM and NR; acc. for mercaptan-modified CR; antidegr. for NR-latex and for thermoplastic SBR.
△ Acc. 22
△ Antidegr. 109

**36** 1. N,N'-Diphenylguanidine
2. $HN{=}C(NH.C_6H_5)_2$
3. Accicure DPG — Alkali
Denax — Dimitrova
Denax DPG — Vychodočeské
DPG — Cyanamid
DPG — Anchor
DPG — Monsanto
DPG — Rhône Poulenc
Eveite D — ACNA Montecatini
Nocceler D — Ouchi Shinko
Pennac DPG — Pennwalt
Sanceler D — Sanshin
Soxinol D — Sumitomo
Vulcafor DPG (surface coated) — ICI
Vulcafor DPG — ICI (India)
Vulkacit D — Bayer
4. White crystalline non-hygroscopic powder or paste. M.W. 211.26; M.R. 144–146 °C.; S.G. 1.19.
Medium acc. for use with thiazoles and sulphenamides.
△ Acc. 31

**37** 1. Sym. Diphenyl-thiourea
syn. Thiocarbanilide

2. $C_6H_5.NH.CS.NH.C_6H_5$
3. A-L Thiocarbanilide — Monsanto
   DPTU — Prochim
   Eveite TC — ACNA Montecatini
   Nocceler C — Ouchi Shinko
   Soxinol C — Sumitomo
   Stabilisator C (formerly Vulkacit CA) — Bayer
   Vulcafor TC — ICI
   — — Degussa
4. Cream-white powder. M.W. 228; M.P. 149 °C. min.; S.G. 1.31.
   Non-staining sec. acc. for CR, EPDM.
   △ Acc. 25

**38**

1. N,N'-Di-ortho-tolylguanidine
2. $2\text{–}CH_3.C_6H_4.NH.C(:NH)NH.C_6H_4.CH_3\text{–}2$
3. DOTG — Anchor
   DOTG — Cyanamid
   DOTG — Du Pont
   DOTG — Rhône Poulenc
   Eveite DOTG — ACNA Montecatini
   Nocceler DT — Ouchi Shinko
   Soxinol DT — Sumitomo
   Vulcafor DOTG — ICI
   Vulcafor DOTG (surface coated) — ICI
   Vulkacit DOTG — Bayer
   Vulkacit DOTG/C (surface coated) — Bayer
4. White powder. M.W. 239; M.R. 167–173 °C.; S.G. 1.19.
   Act. for acidic and neutral acc.; slow-curing acc. for NR, SBR, NBR.
   △ Acc. 32

**39**

1. Polyoxyethyleneglycol
   syn. Polyethyleneglycol
2. $H\text{–}(O\text{–}CH_2\text{–}CH_2)_n\text{–}OH$

3. Glicogum 4000 — Bozzetto a
   PEG — Shell b
   grades 200, 300, 400, 555M, 600, 800, 1000, 1500, 4000, 4000F, 4000P, 6000
4. a. White flakes. M.R. 54–57 °C.; S.G. 1.12.
   b. The numerical suffix is an indication of the average molecular weight.
   PEG 200, 300, 400 colourless liquids. PEG 555M, 600 soft white materials. PEG 800, 1000, 1500, 4000, 6000 white waxy solids.
   Act. for NR, SBR.

**40**
1. ortho-Tolylbiguanide
2. $H_2N{-}C(:NH){-}NH{-}C(:NH){-}C_6H_4.CH_3$
3. Accelerator 80 — Du Pont
   Eveite 1000 — ACNA Montecatini
   Nocceler BG — Ouchi Shinko
   Vulkacit 1000 — Bayer
   Vulkacit 1000/C (surface coated) — Bayer
4. White powder. M.W. 191.24; M.P. 140 °C.; S.G. 1.2.
   Act. for zinc dithiocarbamates, thiuram-, mercapto- and sulphenamide-type acc.; acc. for NR, IR, BR, SBR, NBR.
   △ Acc. 34

**41**
1. Triallylcyanurate
2. $(CH_2{:}CHCH_2{-}OC{:}N{-})_3$ (with a vertical bond from the C)
3. Aktivator OC — Bayer
   Aktivator OC — Degussa
4. Viscous, pasty or crystalline, depending on the temperature. M.W. 249; M.P. 29 °C.; B.P. 149–150 °C. (at 2 mm Hg); S.G. 1.14.
   Act. for peroxide-vulcanization.

**42**
1. Modified urea
2. —
3. Activator 1203 — Anchor a
   Activator DN — Du Pont b

| | | | |
|---|---|---|---|
| | | Attivante 4030 | Bozzetto a |
| | | Attivante 4030 pasta | Bozzetto a |
| | | BIK | Naugatuck SpA b |
| | | BIK | Uniroyal b |
| | | BK | Ouchi Shinko b |
| | | RIA NC | Nat. Polychemicals a |

4. White powders, S.G. ca. 1.3.
   a. act. for thiazole-, thiuram- and sulphenamide type acc.
   b. act. for nitrogeneous blowing agents.

**43** 1. Zinc salts of a mixture of fatty acids in which lauric acid predominates.

2. $$CH_3-(CH_2)_{10}-\overset{\overset{\displaystyle O}{\|}}{C}-O-Zn-O-\overset{\overset{\displaystyle O}{\|}}{C}-(CH_2)_{10}-CH_3$$

| | |
|---|---|
| 3. Laurex | Naugatuck SpA |
| Laurex | Uniroyal |

4. Yellowish waxy powder. M.P. 95–105 °C.; S.G. 1.15.
   Non-staining act. for NR, SBR, NBR, CR.

**44** 1. Composition undisclosed

2. —

| | |
|---|---|
| 3. Additiv-Paste 1100 | Rhein Chemie |
| Additiv-Paste 1600 | Rhein Chemie |
| Aktiplast | Rhein Chemie |
| Aktiplast T | Rhein Chemie |
| Aktivator 3555 | Rhein Chemie |
| Aktivator B | Rhein Chemie |
| Kenmix | Kenrich a |
| Ridacto | Kenrich |
| Ridacto 75 (75 % dispersion) | Kenrich |
| Ritardante 01 | Bozzetto |
| Silacto | Kenrich |
| Silacto 75 (75 % dispersion) | Kenrich |

4. a. △ Acc. 141
   △ Antidegr. 125
   △ Vulc. 54

# Antidegradants

## Abbreviations

| | |
|---|---|
| Antioxidant | oxi |
| Antiozonant | ozo |
| Antiflexcracking-agent | flex |
| Heat-stabilizer | heat |
| Metalpoison-inhibitor | inh |

## INORGANIC

**1**

1. Magnesium oxide
2. $MgO$
3. Kenmag — Kenrich a
   Maglite — Merck b
     grades D, L, K, M, Y
   RC Granulat MgO — Rhein Chemie c
     (80 % MgO, 20 % saturated hydrocarbons and dispersion agents)
   Scorchguard — Anchor d
     grades C3, O, W
   Scorchguard O — Newalls e
   Struktol — Schill & Seilacher f
     grades WB 900, WB 902 (coated)
4. M.W. 40.32.
   a. S.G. 2.02. Act. for CR.
   b. White powders. S.G. 3.3–3.5.
      Antidegr. for CR, CSM, chlorobutyl, SBR, fluoro-elastomers; act. for SBR, and fluoro-elastomers; vulc. for CR, CSM.
   c. S.G. 2.06.
   d. Grade C3, powder, heavy calcined MgO. Grade O, putty, light calcined MgO. Grade W, powder, light calcined MgO.

e. Dispersion. S.G. 2.08.
f. Act. for CR and CSM mixtures.
△ Act. 2
△ Vulc. 2

## ORGANIC

### Hydrocarbons and waxes

2
1. Waxes and paraffinic products
2. —
3. 

| Product | Supplier |
|---|---|
| AC Polyethylene grades 6, 6A, 7, 8, 8A, 615, 617, 617A, x1702, G 201. | Allied |
| Antilux grades 540, 550, 600, 654, AO, AOL, L. | Rhein Chemie |
| Antisun | Ross |
| Aristowax 165 | Ross |
| Controzon grades ASM, Plus, T, W. | Grandel |
| Heliozone | Du Pont |
| Heliozone Special | Du Pont |
| Lichtschutzwachs grades 85R, 102, 116, 38, 335M. | Schlickum |
| Ozonschutzwachs 110 | Bayer |
| Ozonschutzwachs 111 | Bayer |
| Ozonschutzwachs 110 | Rhein Chemie |
| Ozonschutzwachs 111 | Rhein Chemie |
| Ozonschutzwachs DOG | Grandel |
| RC Lichtschutzmittel 520 | Rhein Chemie |
| Shellwax grades 100, 200, 400. | Shell |
| Sunnoc | Ouchi Shinko |
| Sunnoc N | Ouchi Shinko |
| Sunproof grades 100, Regular, Improved, Super. | Naugatuck SpA |

| | | | |
|---|---|---|---|
| | | Sunproof<br>grades Extra, Super, Improved,<br>Regular, Junior, 713. | Uniroyal |
| | | Sunproofing Wax<br>grades 1343, 3920. | Ross |
| | 4. | — | |

**Phenols**

| | | | |
|---|---|---|---|
| **3** | 1. | Alkylated p-cresol | |
| | 2. | — | |
| | 3. | Naugawhite 434 | Uniroyal |
| | 4. | Viscous yellow liquid. S.G. 1.08.<br>Oxi for dry NR and for NR- and SBR latex. | |
| **4** | 1. | Alkylated bisphenol | |
| | 2. | — | |
| | 3. | Antigene NW | Sumitomo a |
| | | Cyanox 53<br>(on inert carrier) | Cyanamid b |
| | | Naugawhite | Naugatuck SpA a |
| | | Naugawhite powder<br>(on inert carrier) | Naugatuck SpA b |
| | | Naugawhite | Rubber Regenerating a |
| | | Naugawhite | Uniroyal a |
| | | Naugawhite powder<br>(on inert carrier) | Uniroyal b |
| | 4. | a. Slightly viscous liquid. S.G. 0.96.<br>b. White powder. S.G. 1.19.<br>Non-staining oxi for NR, SBR, CR, IR, NBR, BR. | |
| **5** | 1. | Alkylated phenols blends | |
| | 2. | — | |
| | 3. | Antioxidant KSM | Bayer |
| | | Antioxidant KSM-EM-33 | Bayer |
| | | Arrconox AHT | Rubber Regenerating |

| | | | |
|---|---|---|---|
| | | Arrconox AHT powder (on inert carrier) | Rubber Regenerating |
| | | Cyanox LF | Cyanamid |
| | | Cyanox LF | Anchor |
| | | Oxystop grades 320, 330. | Arwal |
| | | Wingstay T | Goodyear |
| | 4. | Non-staining powders and liquids. Oxi for natural and synthetic rubber. | |
| **6** | 1. | Alkylated styrenated phenol | |
| | 2. | — | |
| | 3. | Wingstay V | Goodyear |
| | 4. | Amber liquid. S.G. 1.00. FDA appr. Non-staining oxi for natural and synthetic rubber. Also for latex. | |
| **7** | 1. | Alkylated thio-bisphenol | |
| | 2. | — | |
| | 3. | Antioxidant 423 (form. trade-name VBUI) | Uniroyal |
| | 4. | White crystalline solid. M.P. 115–117 °C.; S.G. 1.07. Non-staining oxi for rubber. | |
| **8** | 1. | Alkyl- and aryl-substituted phenol blend | |
| | 2. | — | |
| | 3. | Antioxidant DS (form. Antioxidant KA 9013) | Bayer a |
| | | Antioxidant DS/F | Bayer b |
| | | Antioxidant TSP | Bayer c |
| | 4. | a. Yellow-reddish viscous liquid. B.P. 150 °C. min. (10 Torr); S.G. 0.915. Non-staining oxi, heat, flex for NR, SBR, IR, BR, NBR, CR. | |
| | | b. White powder with inorganic filler, 50 % active ingredient. Non-staining oxi, heat, flex for mixtures of NR with SBR, IR, BR, NBR, CR. | |
| | | c. Viscous orange-red liquid. B.P. 170–180 °C. (10 Torr). | |
| **9** | 1. | Alkylphenol sulphide | |
| | 2. | — | |

3. Rhenadox APS — Rhein Chemie
4. White crystalline powder. M.P. 72 °C. min.; S.G. 1.06.
   Non-staining oxi, heat.

**10**
1. 1,1-Bis-(4-hydroxy phenyl) cyclohexane
2. $CH_2$–$CH_2$, $H_2C$, C, OH, OH, $CH_2$–$CH_2$
3. Antigene W — Sumitomo
4. White powder. M.W. 268; M.P. 175 °C. min.
   Non-staining oxi, heat for NR, SBR.

**11**
1. 6-tert Butyl, 2,4-dimethyl phenol
2. OH, $(CH_3)_3C$—, —$CH_3$, $CH_3$
3. Antioxidant 624 — Raschig
4. Yellow liquid. M.W. 178.26; B.P. 250 °C.; S.G. 0.95–0.96.
   Oxi.

**12**
1. 6-tert Butyl, 3-methyl phenol derivatives
2. —
3. Antigene WL–L — Sumitomo
   Antigene WL–O — Sumitomo
4. Reddish-brown tacky liquid.
   Non-staining oxi for NR and for synthetic rubber.

**13**
1. 2,2'-Butylidene-bis-(6-tert butyl p-cresol)
2. $CH_3$, $CH_3$, $(CH_3)_3C$—, —$C_4H_8$—, —$C(CH_3)_3$, OH, OH
3. Anullex PBH — Pearson

4. Off-white crystalline powder. M.W. 382.24; M.P. 127 °C.; S.G. 0.56.
   Non-staining oxi for dry rubber and latex.

**14**
1. 4,4'-Butylidene-bis-(2-tert butyl 5-methylphenol)
   syn. 4,4'-Butylidene-bis-(6-tert butyl m-cresol)
2. $(CH_3)_3C$—, HO— (ring) —$C_4H_8$— (ring) —$C(CH_3)_3$, —OH; $CH_3$ $H_3C$
3.

| | |
|---|---|
| Antigene BBM | Sumitomo |
| Anullex PBA 15 | Pearson |
| Santowhite Powder | Monsanto |
| Sumilizer BBM | Sumitomo |

4. White crystalline powder. M.W. 382.56; M.P. 209 °C.; S.G. 1.03–1.09.
   Non-staining oxi for natural and synthetic rubber, also for latex. FDA appr.

**15**
1. 2,6-Di-tert butyl-p-cresol
   syn. 2,6-Di-tert butyl-4-methyl phenol
   2,6-Di-tert butyl hydroxy-toluene
2. $C_6H_2(OH)[(CH_3)_3C-]_2(CH_3)$–1–2,6–4
3.

| | |
|---|---|
| Antigene BHT | Sumitomo |
| Antioxidant 4 | Dimitrova |
| Antioxidant DBPC grades AT 1, AP 3, AP 4, AB 2. | Organo Synthèse |
| Antioxidant KB | Bayer |
| Anullex BHT | Pearson |
| Anullex BHT techn. grade 1 | Pearson |
| B.H.T. | Bennett |
| B.H.T. | Raschig |
| Bisoxol 220 | CdF |
| Catalin Antioxidant grades CAO-1, CAO-3. | Ashland |
| DTBP | Raschig |
| Imbutol | Metallgesellschaft |
| Imbutol E (50 % aquous solution) | Metallgesellschaft |

| | |
|---|---|
| Ionol CP | Shell |
| Narox | Naugatuck SpA |
| Naugard BHT | Uniroyal |
| Nocrac 200 | Ouchi Shinko |
| Nonox TBC (only available in Great Britain) | ICI |
| Rhenadox DB | Rhein Chemie |
| Tenamene 3 | Eastman |

4. White to yellow flakes. M.W. 220; M.P. 68.9 °C.; S.G. 1.03.
   Non-staining oxi, heat, flex for NR, IR. Also for latex. In combination with waxes ozo.

**16**

1. 4,4'-Thio-bis-(6-tert butyl-m-cresol)
   syn. 4,4'-Thio-bis-(6-tert butyl-3 methyl phenol)
   4,4'-Thio-bis-(2-tert butyl-5 methyl phenol)
   Di-(3-tert butyl-4-hydroxy-6-methyl phenyl) sulphide
2. $C(CH_3)_3$ $C(CH_3)_3$
   HO—(ring)—S—(ring)—OH
   $CH_3$ $CH_3$
3.

| | |
|---|---|
| Antigene WX | Sumitomo |
| Antigene WX–R (pure grade) | Sumitomo |
| Antioxidant TBM-6 T grade | Organo Synthèse |
| Anullex PSA 10 | Pearson |
| Nocrac 300 | Ouchi Shinko |
| Santowhite Crystals | Monsanto |
| Sumilizer WX | Sumitomo |
| Sumilizer WX–R (pure grade) | Sumitomo |

4. White to greyish-white powder. M.W. 358.55; M.P. 150 °C. min.; S.G. 1.06–1.12.
   Non-staining oxi and heat for natural and synthetic rubber, especially for CR, also for latex. FDA appr.

**17**

1. 3,5-Di-tert butyl-4-hydroxy-toluene

2. 
```
                 OH
                 |
(CH3)2C—[benzene ring]—C(CH3)2
                 |
                CH3
```

3. Antigene BHT — Sumitomo
4. White powder or flakes. M.W. 220; M.P. 69 °C. min.
   Non-staining oxi for NR, NBR, IR, PE, also for latex; inh for NR.

**18**
1. 4,4'-Dihydroxydiphenyl
2. $HOC_6H_4C_6H_4OH$
3. Antioxidant DOD — Bayer
4. Gray powder. M.W. 186; M.P. 260 °C.; S.G. 1.37.
   Non-staining oxi, heat for NR dry rubber, and in combination with di-beta-naphthyl-p-phenylenediamine for NR latex.

**19**
1. N,N'-Disalicylidene-1,2-propanediamine
2. 
```
          H     H  CH3    H
          |     |   |     |
[C6H4]—C=N—C—C—N=C—[C6H4]
  —OH           |   |      HO—
                H   H
```

3. Copper Inhibitor 50 — Du Pont
4. Reddish-brown liquid. M.W. 270; Flash-Point 36 °C.; S.G. 0.99.
   Inh for NR, SBR, CR. Also for latex.

**20**
1. 1,6-Hexanediol bis-[3-(3',5'-di-tert butyl-4'-hydroxyphenyl) propionate]
2. 
```
   C(CH3)3            O               O             C(CH3)3
                      ||              ||
HO—[ring]—CH2-CH2-C-O-(CH2)6-O-C-CH2-CH2—[ring]—OH

   C(CH3)3                                          C(CH3)3
```

3. Irganox 259 — Ciba Geigy
4. White crystalline powder. M.W. 638.9; M.P. 104–109 °C.
   Oxi.

**21**
1. hindered Bisphenol
2. —

3. Antioxidant NKF — Bayer
4. White crystalline powder. M.P. 155 °C. min.; S.G. 1.12.
   Oxi, heat, flex. Inh in combination with Antioxidant MB.
   (2-Mercapto-benzimidazole).

**22**

1. hindered Phenols
2. —
3. 

| | |
|---|---|
| Agerite Geltrol | Vanderbilt a |
| Antioxidant NV 1 | Bayer a |
| Antioxidant 431 | Uniroyal a |
| Antioxidant 555 | Pitt Consol a |
| Endox 11T | Akron b |
| Nevastain Powder 2170 | Neville b |
| Santowhite 54 | Monsanto a |
| Santowhite 54–S (on inert carrier) | Monsanto b |
| Wingstay L | Goodyear b |
| Zalba Special | Du Pont b |

4. a. liquid
   b. powder.
   Oxi, flex.

**23**

1. 2,2'-Methylene-bis-(4-ethyl-6-tert butylphenol)
2. $CH_2[-(2)-C_6H_2(OH)C_2H_5\{C(CH_3)_3\}-1,4,6]_2$
3. 

| | |
|---|---|
| Antioxidant 425 | Cyanamid |
| Antioxidant 425 | Anchor |
| Antioxidant TBE 9 | Organo Synthèse |
| Endox 22 | Akron |
| Naruxol 25 | Naugatuck SpA |

4. Cream to white powder. M.W. 368.54; M.R. 119–125 °C.; S.G. 1.10.
   Oxi, heat, flex for NR, SBR, NBR, CR. Also for latex.

**24**

1. 2,2'-Methylene-bis-(4-methyl-6-tert butylphenol)
2. $(CH_3)_3C$ OH HO $C(CH_3)_3$

   $CH_2$

   $H_3C$ $CH_3$

3. Antigene MDP — Sumitomo
   Antioxidant 2246 — Cyanamid
   Antioxidant 2246 — Anchor
   Antioxidant BKF — Bayer
   Bisoxol O — CdF
   Catalin Antioxidant grades CAO-5, CAO-14. — Ashland
   Endox 21T — Akron
   Naruxol 15 — Naugatuck SpA
   Nocrac NS-6 — Ouchi Shinko
   Sumilizer MDP — Sumitomo
   Synox 5LT — Neville Synthèse
   Synox 5P (pure grade) — Neville Synthèse
4. Colourless to light-cream crystalline powder. M.W. 340.51; M.P. 120 °C, min.; S.G. 1.08.
   Non-staining oxi, heat, flex, inh for NR, BR, SBR, NBR. Also for latex.

**25**
1. 2,2'-Methylene-bis-(dimethyl-4,6-phenol)
2. OH OH
   $CH_3$— —$CH_2$— —$CH_3$
   $CH_3$ $CH_3$
3. Permanax 28 "H.V." — Rhône Poulenc
4. Oiled powder, cream coloured. M.W. 256; M.P. 120 °C.; S.G. 1.1.
   Oxi, heat, flex for NR, SBR. Also for latex.

**26**
1. 2,2-Methylene-bis-[6-(alpha methyl cyclohexyl)-p-cresol]
2. 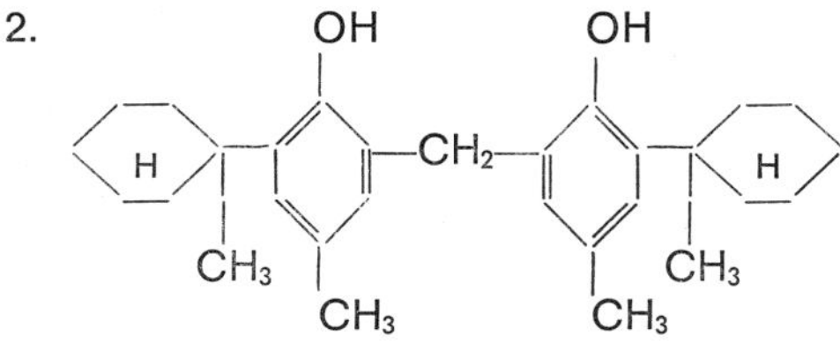

3. Nonox WSP — ICI
   Nonox WSP — ICI America

4. Off-white crystalline powder. M.W. 420; M.P. 130 °C.; S.G. 1.17.
Non-staining oxi for synthetic rubber and for polyolefins.

**27**
1. 2,2'-Methylene-bis-(4-methyl-6-cyclohexylphenol)
2. $CH_2$–[–(2)–$C_6H_2$.(OH).$CH_3$.$C_6H_{13}$–1,4,6]$_2$
3. Antioxidant ZKF — Bayer
4. White crystalline powder. M.W. 380; M.P. 118 °C.; S.G. 1.08.
Oxi, heat, flex, inh for NR, IR, BR, SBR, NBR.

**28**
1. 2,2'-Methylene-bis-(tert nonyl-cresol)
2. $CH_2$–[–(2)–$C_6H_2$.(OH).$CH_3$.$C_9H_{19}$]$_2$
3. Anox G 1 — Bozzetto a
Anox G 1 polvere — Bozzetto b
(absorbed on molecular sieve)
4. a. Viscous amber-coloured liquid. S.G. 0.96.
b. Light brown powder. S.G. 1.01.
Non-staining oxi, heat for NR, SBR, CR.

**29**
1. Aryl-alkyl phenol blends
2. —
3. Permanax 24 — Rhône Poulenc
4. Yellow liquid. S.G. 1.079 and cream powder (60 % Permanax 24 and 40 % mineral powder).
Non-staining oxi, heat for NR, SBR, NBR, IR, CR, BR.

**30**
1. Phenol-derivatives blends
2. —
3. Antioxidant MP — Anchor a
Nonox HO — ICI b
4. a. Amber liquid. B.P. 250 °C.; S.G. 0.95.
b. Amber liquid. S.G. 0.94.
Oxi, flex.

**31**
1. Polybutylated bisphenol-A blend
2. —
3. Agerite Superlite — Vanderbilt a
Agerite Superlite solid — Vanderbilt b
(70 % active ingredient)

4. a. Amber liquid. S.G. 0.965.
   b. Gray-tan solid. S.G. 1.26.
   Oxi.

**32** 1. Nickel salt of 3,5-di-tert butyl-4-hydroxybenzyl-phosphonic acid-monoethylester

2. $\left[ HO-C_6H_2(C(CH_3)_3)_2-CH_2-\overset{O \uparrow}{P}(OC_2H_5)-O^{\ominus} \right]_2 Ni^{2\oplus}$

3. Irgastab 2002 Ciba Geigy
4. Pale-yellow powder. M.W. 713.47.
   Oxi, light stabilizer.

**33** 1. Octadecyl-3-(3',5'-di-tert butyl-4'-hydroxyphenyl) propionate

2. $HO-C_6H_2(C(CH_3)_3)_2-CH_2CH_2-CO-O-C_{18}H_{37}$

3. Irganox 1076 Ciba Geigy
4. White crystalline powder. M.W. 530.9; M.P. 49–54 °C.
   Oxi.

**34** 1. Pentaerythritol-tetrakis-[3-(3',5'-di-tert butyl-4'-hydroxyphenyl) propionate]

2. $\left[ HO-C_6H_2(C(CH_3)_3)_2-CH_2CH_2-CO-O-CH_2- \right]_4 C$

3. Irganox 1010 Ciba Geigy
4. White crystalline powder. M.W. 1177.7; M.P. 110–115 °C.
   Oxi.

**35** 1. 1,1-Bis-(4-hydroxy phenyl) cyclohexane - organic amine reaction product

2. —
3. Antigene WA — Sumitomo
4. Light brown powder. M.P. 80 °C. min.
   Non-staining oxi, ozo, heat for NR, SBR, NBR, CR.

**36**
1. 6-tert Butyl-m-cresol - sulphur dichloride reaction product
2. —
3. Santowhite MK — Monsanto
4. Dark-brown viscous liquid. M.P. 20–35 °C.; S.G. 1.03–1.09.
   Non-staining oxi for dry rubber and latex. FDA appr.

**37**
1. Styrenated phenols
2. OH-substituted phenyl ring bonded to $\left[ C(H)(CH_3)-C_6H_5 \right]_n$

   (structure: OH, H, C, $CH_3$, n)
3.

| | |
|---|---|
| Accinox SP | Alkali a |
| Agerite SPAR | Vanderbilt |
| Anox G2 | Bozzetto |
| Anox G2 Polvere (absorbed on molecular sieves) | Bozzetto |
| Antigene S | Sumitomo |
| Antioxidant SP | Anchor |
| Arrconox SP | Rubber Regenerating |
| Arrconox SP Powder (dispersed in an inert powder base) | Rubber Regenerating |
| Montaclere | Monsanto |
| Montaclere SE (self-emulsifying grade) | Monsanto |
| Nocrac SP | Ouchi Shinko |
| Nonox SP | ICI a |
| Nonox SP | ICI (India) a |
| Stabilite SP | Reichhold |
| Wingstay S | Goodyear |

4. Light-yellowish transparent gluey liquid.
   Non-staining oxi, ozo, heat for NR, SBR, NBR, CR. FDA appr.
   a. Mixture of styrenated phenols.

**38** 1. 4,4'-Thio-bis-(di-sec amyl phenol)
2. $C_{32}H_{50}O_2S$
3. Santowhite L — Monsanto
4. Dark viscous liquid. M.W. 508.8; S.G. 0.94–1.04.
Non-staining oxi for latex. FDA appr.

**39** 1. 1,3,5-Trimethyl-2,4,6-tris-(3,5-di tert butyl-4-hydroxybenzyl) benzene
2.
$C(H_3)_3$ $C(CH_3)_3$
HO OH
$(CH_3)_3C$ $CH_3$ $C(CH_3)_3$
$H_2C$ $CH_2$
$H_3C$ $CH_3$
$CH_2$
$(CH_3)_3C$ $C(CH_3)_3$
OH
3. Ionox 330 — Shell
4. Near-white crystalline solid. M.W. 768; M.P. 244 °C.
Oxi.

## Derivatives of aniline and of substituted aniline

**40** 1. Acetaldehyde-aniline condensation product
2. —
3. Crylene (mixture with stearic acid 67 : 33) — Uniroyal a
Nocceler K — Ouchi Shinko b
VGB — Uniroyal c
Vulcafor RN — ICI †
4. a. Thick brown paste. S.G. 1.014.
b. Reddish-brown powder. M.P. 55 °C.

c. Brown resinous powder. M.R. 60–80 °C,; S.G. 1.15.
Staining oxi, heat for NR; acc. for NR.
△ Acc. 11

**41** 1. Butyraldehyde - aniline condensation product
2. —
3. | | |
|---|---|
| Accelerator 21 | Anchor |
| Antox Special | Du Pont |
| Beutene | Uniroyal |
| Butanyl-1 | Ticino |
| Nocceler 8 | Ouchi Shinko |
| Rapid Accelerator 300A | Rhône Poulenc |
| Vulcafor BA | ICI |

4. Orange-red oily liquid. S.G. 0.94–1.02.
Antidegr. for CR; semi-ultra acc. for Neoprene W and hydrocarbon rubber; act. for thiazoles, thiurams and guanidines.
△ Acc. 14
△ Act. 11

**42** 1. 4,4'-Diamino diphenyl methane
2. $CH_2(C_6H_4.NH_2)_2$
3. | | |
|---|---|
| Robac 4.4 | Robinson a |
| Tonox | Uniroyal b |

4. a. Light brown powder. M.W. 198.3; M.R. 75–85 °C.
b. Brown waxy lump. S.G. 1.18.
Antifrosting agent for NR; acc. for CR; ret. for IIR.
△ Acc. 15
△ Ret. 2

**43** 1. N,N-Di-(N'-ethylidene-anilino) aminobenzene
2.
$$\begin{array}{c} \qquad\qquad\qquad C_6H_5 \\ \qquad\qquad\qquad | \\ C_6H_5\text{–}NH\text{–}HC\text{–}N\text{–}CH\text{–}NH\text{–}C_6H_5 \\ \qquad | \qquad\quad | \\ \qquad CH_3 \quad CH_3 \end{array}$$
3. Eveite A — ACNA Montecatini
4. Brown oily liquid. M.W. 331.45; S.G. 1.06–1.07.
Antidegr. and acc.
△ Acc. 16

**44** 1. Arylamines blends
2. —
3. Accinox HFN — Alkali
Anox Beta — Bozzetto
Anox JO — Bozzetto
Nonox HFN — ICI
Nonox HFN — ICI (India)
Nonox HFN — ICIANZ
4. Gray flakes, granules, pellets. S.G. 1.18–1.22.
Staining oxi, heat, flex, for NR, SBR, NBR, CR and for isobutylene-isoprene copolymers.

**45** 1. N-Methyl, N,4-dinitroso-aniline / inert clay blend
2. —
3. Heat Pro — Conestoga
4. Antidegr.

**Derivatives of phenylene-diamine**

**46** 1. Alkyl-aryl p-phenylenediamines
2. —
3. Aceto Ozone — Aceto a
Flexzone 5L — Uniroyal b
Santoflex 134 — Monsanto c
Santoflex 134 SE — Monsanto c
(self-emulsifying grade)
4. a. Blend of alkyl-aryl p-phenylenediamine with waxes. Tan granulated beads. M.P. 107.2 °C.; S.G. 0.98.
Oxi, ozo, flex, inh, heat for NR, SBR.
b. Dark-brown crystalline solid.
Staining oxi, ozo for SBR, IR, BR. (Formerly Antiozonant 437)
c. Dark oil. Crystallization Point $<$18.3 °C.; S.G. 0.993.
Oxi for SBR.

**47** 1. N,N'-Bis-(1-ethyl, 3-methyl pentyl)-p-phenylenediamine
syn. N,N'-Di-(3,5-methyl heptyl)-p-phenylenediamine
2. $(C_2H_5.CHCH_3.CH_2.CHC_2H_5.NH)_2C_6H_4$

3. Antozite 2 — Vanderbilt
   Flexzone 8L — Uniroyal
   Santoflex 17 — Monsanto
   UOP 88 — UOP
4. Dark red-brown liquid. M.W. 332.56; S.G. 0.87–0.93.
   Staining ozo, flex, for natural and synthetic rubber.

**48**
1. N,N'-Bis-(1-methyl heptyl)-p-phenylenediamine
   syn. N,N'-Di-(2-octyl)-p-phenylenediamine
2. $C_6H_{13}-CH(CH_3)-NH-C_6H_4-NH-CH(CH_3)-C_6H_{13}$
3. Antozite 1 — Vanderbilt
   $ANTO_3$"B" — Pennwalt
   $ANTO_3$"D" — Pennwalt
   (isomeric)
   Santoflex 217 — Monsanto
   UOP 288 — UOP
4. Dark red-brown liquid. M.W. 332.56; S.G. 0.87–0.93.
   Staining oxi for NR.

**49**
1. N,N'-Bis-(1,4-dimethyl pentyl)-p-phenylenediamine
2. $CH_3-CH(CH_3)-CH_2-CH_2-CH(CH_3)-NH-C_6H_4-NH-CH(CH_3)-CH_2-CH_2-CH(CH_3)-CH_3$
3. Antioxidant 4030 — Bayer
   Antozite MPD — Vanderbilt
   Cyzone DH — Cyanamid
   Eastozone 33 — Eastman
   Flexzone 4L — Uniroyal
   Santoflex 77 — Monsanto
   UOP 788 — UOP
4. Reddish-brown liquid. M.W. 304.51; S.G. 0.894–0.906.
   Staining general purpose ozo.

**50**
1. Diaryl-p-phenylenediamine (mixed)
2. —

3. Akroflex AZ — Du Pont a
   Antigene DTP — Sumitomo b
   Wingstay 100 — Goodyear c
   Wingstay 200 — Goodyear d
4. a. Brown powder. M.P. 103 °C.; S.G. 1.27.
   Staining ozo for SBR, CR.
   b. Black powder.
   Oxi, ozo, heat, flex for NR, SBR, NBR.
   c. Blue-brown solid. M.P. 90–105 °C.; S.G. 1.20.
   Oxi, ozo, flex.
   d. Brown-black semi-solid. M.P. 50–60 °C.; S.G. 1.20.
   Oxi, ozo, flex.

**51**
1. N,N'-Di-cyclohexyl-p-phenylenediamine
2. $C_6H_{11}.NH.C_6H_4.NH.C_6H_{11}$
3. UOP 26 — UOP
4. Brown flakes. M.W. 272.4; M.P. 102–108 °C.; S.G. 0.59.
   Staining ozo, flex for NR, NBR, CR, IR, BR.

**52**
1. N,N'-Di-heptyl-p-phenylenediamine
2. $C_7H_{15}.NH.C_6H_4.NH.C_7H_{15}$
3. ANTO$_3$"G" — Pennwalt
4. Reddish-brown liquid. M.W. 304; S.G. 0.898.
   Staining ozo for NR, SBR.

**53**
1. N-(1,3-Dimethyl butyl)-N'-phenyl-p-phenylenediamine
2. $CH_3.CHCH_3.CH_2.CHCH_3.NH.C_6H_4.NH.C_6H_5$
3. Antioxidant 4020 — Bayer
   Antozite 67 — Vanderbilt
   Antozite 67 S — Vanderbilt
   (50 % active ingredient)
   Flexzone 7L — Uniroyal
   Nonox ZC — ICI
   Santoflex 13 — Monsanto
   Santoflex 13 S — Monsanto
   (Santoflex / Carbon black 1 : 1)
   UOP 588 — UOP

4. Dark coloured crystals or flakes. M.W. 268.39; M.R. 40–44 °C.; S.G. 0.9–1.1.
   Staining general purpose oxi, ozo, flex, inh.

**54**
1. N,N'-Di-beta-naphthyl-p-phenylenediamine
2. $C_{10}H_7$—NH—$C_6H_4$—NH—$C_{10}H_7$ (structural formula)
3. Aceto DIPP — Aceto
   Agerite White — Vanderbilt
   Agerite White — Anchor
   (Antioxidant 123 outside U.K.)
   Antigene F — Sumitomo
   Antioxidant DNP — Bayer
   Antivecchiante DNP — ACNA Montecatini
   Nocrac White — Ouchi Shinko
   Nonox CI — ICI
   Nonox CI — ICIANZ
   Santowhite CI — Monsanto
4. Grey powder. M.W. 360; M.P. 230 °C.; S.G. 1.28.
   Staining oxi, heat, inh for NR, SBR, NBR. Also for latex.

**55**
1. N,N'-Diphenyl-p-phenylenediamine / phenyl-alpha-naphthylamine blend
2. —
3. Akroflex C — Du Pont a
   (65 : 35)
   Nocrac 500 — Ouchi Shinko
4. a. Dark-gray pellets. M.P. 75 °C.; S.G. 1.23.
   Staining oxi, ozo, flex, heat for NR, SBR, CR, IIR.

**56**
1. N,N'-Diphenyl-p-phenylenediamine
2. $C_6H_5.NH.C_6H_4.NH.C_6H_5$
3. Agerite DPPD — Vanderbilt
   Antigene P — Sumitomo
   Antioxidant DPPD — Anchor
   DPPD — Monsanto
   Inibitore OB — ACNA Montecatini a
   J-Z-F — Uniroyal

| | |
|---|---|
| Nocrac DP | Ouchi Shinko |
| Nonox DPPD | ICI |
| Permanax 18 | Rhône Poulenc |

4. Light gray powder. M.W. 260; M.P. 140 °C.; S.G. 1.22.
   a. M.P. 125–130 °C.
   Staining oxi, ozo, flex, heat for NR, SBR, NBR, IR, BR, CR.

**57**
1. N-Isopropyl-N'-phenyl-p-phenylenediamine
   syn. 4-Isopropylamino-diphenylamine
2. $C_3H_7.NH.C_6H_4.NH.C_6H_5$
3. Trade names:

| | |
|---|---|
| Antigene 3C | Sumitomo |
| Antioxidant 4010 NA | Bayer |
| Antiozonant IP | Anchor |
| Eastozone 34 | Eastman a |
| Flexzone 3C | Uniroyal |
| Nocrac 810–NA | Ouchi Shinko |
| Nonox ZA | ICI |
| Permanax 115 | Rhône Poulenc |
| RC Granulat IPPD (80 % IPPD, 20 % saturated hydrocarbons and dispersing agents) | Rhein Chemie |
| Santoflex IP | Monsanto |

4. Brown flakes. M.W. 226.3; M.P. 74 °C. min.; S.G. 1.01–1.07.
   a. Liquid. S.G. 0.90; Flash Point 203 °C.
   Staining oxi, ozo, heat, flex, inh for NR, SBR, NBR, BR, CR.

**58**
1. N-Methyl-2-pentyl-N'-phenyl-p-phenylenediamine
2. $C_5H_{11}$

   $CH_3$—NH—$C_6H_3$—NH—$C_6H_5$
3. Wingstay 300 — Goodyear
4. Red-black solid. M.W. 268; M.P. 40–50 °C.; S.G. 0.98.

**59**
1. N-Phenyl-N'-cyclohexyl-p-phenylenediamine
2. $C_6H_5.NH.C_6H_4.NH.C_6H_{11}$
3. Antioxidant 4010 — Bayer

Antiozonant CP (U.K. only) — Anchor
Flexzone 6H — Uniroyal
UOP 36 — UOP

4. Greyish-white powder. M.W. 267; M.P. 115 °C. min.; S.G. 1.19. Oxi, ozo, heat, flex, inh for NR, SBR, CR.

**60**

1. N-Phenyl-N'-hexyl-p-phenylenediamine
2. $C_6H_5.NH.C_6H_4.NH.C_6H_{13}$
3. ANTO$_3$"E" — Pennwalt
4. Greyish solid. M.W. 268; M.P. 45–50 °C.; S.G. 1.015. Ozo, flex for NR, SBR.

**61**

1. N-Phenyl-N'-2-octyl-p-phenylenediamine
2. $C_6H_5.NH.C_6H_4.NH.C_8H_{17}$
3. ANTO$_3$"F" — Pennwalt
   UOP 688 — UOP
4. Brown viscous liquid. M.W. 296.4. M.P. 10 °C.; B.P. 430 °C.; S.G. 1.003. Ozo, flex for NR, SBR. Staining.

**62**

1. N-Phenyl-N'-(p-toluene-sulphonyl)-p-phenylenediamine
2. 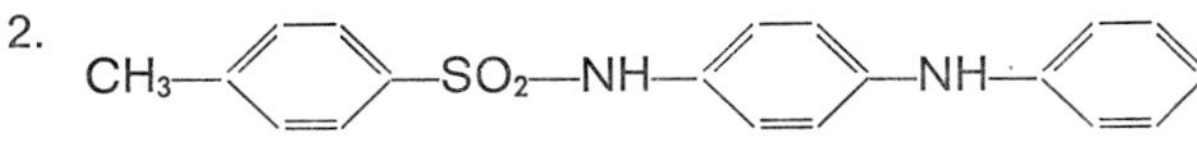

3. Aranox — Uniroyal
4. Gray powder. M.W. 338; M.P. 140 °C.; S.G. 1.35. Oxi, inh for NR, CR.

**63**

1. Diaryl- and alkyl-aryl-p-phenylenediamine blends
2. —
3. Wingstay 250 — Goodyear a
   Wingstay 275 — Goodyear b
4. a. Dark-brown liquid. Liquid at 38 °C.; S.G. 1.06.
   b. Dark-brown liquid containing crystals. Slushy liquid at room temperature. S.G. 1.03.

**64**

1. N,N'-Bis-(1-methyl heptyl)-p-phenylenediamine / N-(1,3-dimethyl butyl)-N'-phenyl-p-phenylenediamine / N-phenyl-(N'-methyl heptyl)-p-phenylenediamine blend
2. —

3. UOP 256 UOP
4. Viscous reddish-brown liquid. Flash-Point 132 °C.; S.G. 0.975. Ozo, flex for natural and synthetic rubber.

**65**
1. N,N'-Diheptyl-p-phenylenediamine / phenyl hexyl-p-phenylenediamine blend
2. —
3. $ANTO_3$"A" Pennwalt
4. Reddish-brown liquid. S.G. 0.964. Staining ozo for NR, SBR.

**66**
1. N,N'-Bis-(1-methyl heptyl)-p-phenylenediamine / N-phenyl-(N'-methyl heptyl)-p-phenylenediamine blend
2. —
3. UOP 62 UOP
4. Viscous reddish-brown liquid. Flash Point 132 °C.; S.G. 0.952. Ozo, flex for natural and synthetic rubber.

**67**
1. N,N'-Bis-(1,4-dimethyl pentyl)-p-phenylenediamine / N-(1,3-dimethyl butyl)-N'-phenyl-p-phenylenediamine blend
2. —
3. UOP 57 UOP
4. Viscous reddish-brown liquid. Flash Point 160 °C.; S.G. 0.9638. Ozo, flex for natural and synthetic rubber.

**68**
1. Dioctyl-p-phenylenediamine / phenyl hexyl-p-phenylenediamine / phenyl octyl-p-phenylenediamine blend
2. —
3. $ANTO_3$"C" Pennwalt
4. Reddish-brown liquid. S.G. 0.975. Non-staining ozo for natural and synthetic rubber.

**Derivatives of diphenylamine**

**69**
1. Alkylated diphenylamine
2. —
3. Pennox A–"S" Powder Pennwalt a
   Wytox ADP Nat. Polychemicals b
   Wytox ADP–X Nat. Polychemicals b

4. a. Grey powder. S.G. 1.26.
   b. Light-tan powder. M.R. 80–88 °C.; S.G. 0.99.
   Oxi for NR.

**70**
1. p-Isopropoxy-diphenylamine
2. $C_6H_5-NH-C_6H_4-O-CH(CH_3)_2$
3. Agerite ISO — Vanderbilt
4. Tan-to-grey flakes. M.W. 227; M.R. 80–86 °C.; S.G. 1.15.
   Oxi.

**71**
1. Diphenylamine-derivatives
2. —
3. Antioxidant DDA — Bayer
   Antioxidant DDA–EM 30 % — Bayer a
   Antioxidant DDA–EM 50 % — Bayer a
4. Viscous brown to reddish liquid. B.P. 300 °C. min.; S.G. 1.08.
   a. Emulsions for latex.
   Slightly staining oxi, flex, heat, inh for NR, IR, SBR, BR.

**72**
1. Diphenylamine - acetone condensation product
2. —
3. Accinox B — Alkali
   Accinox BL — Alkali
   Accinox BLN — Alkali
   Agerite Superflex — Vanderbilt
   Agerite Superflex solid — Vanderbilt
   Aminox — Naugatuck SpA
   Aminox — Uniroyal
   Anoxin — ACNA Montecatini
   Antigene AM — Sumitomo a
   Antioxidant B — Anchor
   BLE 25 — Naugatuck SpA
   BLE 25 — Rubber Regenerating
   BLE 25 — Uniroyal
   BLE 25 GP — Naugatuck SpA
   (BLE 25 70 %, diatomite 25 %)

| | |
|---|---|
| BLE 50 | Uniroyal |
| (BLE 50 %, carbon black 50 %) | |
| Cyanaflex 50 | Cyanamid |
| (absorbed on carbon black) | |
| Cyanaflex 100 | Cyanamid |
| Neozone L | Du Pont |
| Nocrac B | Ouchi Shinko |
| Nonox B | ICI |
| Nonox B | ICI (India) |
| Nonox BL | ICI |
| Nonox BLB | ICI |
| (absorbed on carbon black) | |
| Nonox BLN | ICI |
| Nonox BLW | ICI |
| (absorbed on silica carrier) | |
| Permanax 47 | Rhône Poulenc |

4. Brown powder, liquid, pellets and rods. M.P. 75–95 °C.; S.G. liquid 1.09; S.G. powder 1.14.
   a. M.P. 190 °C. min.
   Staining ozo, heat, flex for NR, SBR, NBR, CR.

**73**
1. Diphenylamine - diisobutylene reaction product
2. —
3. Arrconox DNL — Rubber Regenerating a
   Octamine — Rubber Regenerating
   Octamine — Naugatuck SpA
   Octamine — Uniroyal
4. Light-brown granular waxy solid. M.R. 77–85 °C.; S.G. 0.99.
   a. Dark liquid. S.G. 0.95.
   Staining oxi, heat, flex for NR, SBR, CR, NBR.

**74**
1. Diphenylamine - olefin reaction product
2. —
3. Anox NS — Bozzetto
4. Light-brown granular waxy solid. M.R. 75–88 °C.; S.G. 0.98.
   Oxi, heat, flex for NR, SBR, CR, IIR.

**75**
1. Nonylated diphenylamine
2. —

3. Pennox A — Pennwalt
   Polylite — Uniroyal
4. Dark liquid. S.G. 0.95.
   Oxi for NR, SBR, NBR, IR, CR, BR.

**76**
1. Octylated diphenylamine
2. —
3. Agerite Stalite — Vanderbilt
   Agerite Stalite S — Vanderbilt
   Anox NSL — Bozzetto
   Antioxidant OCD — Bayer a
   Antox N — Du Pont
   Cyanox 8 (flakes and granules) — Cyanamid
   Nonox OD — ICI
4. a. Gray-brown granules. M.P. 88 °C. min.; S.G. 1.00.
   Oxi, flex, heat, inh for NR, CR.
   Oxi, heat for NR, CR.

**77**
1. Alkylated diphenylamines and petroleum wax blend (75 : 25)
2. —
3. Agerite Gel — Vanderbilt
4. Soft, light-tan to brown waxy solid. M.R. 40–50 °C.; S.G. 0.94.
   Oxi.

## Mixtures and derivatives of alpha- and beta-naphthylamine

**78**
1. Aldol-alpha-naphthylamine
2. $N{=}CHCH_2CH(OH)CH_3$

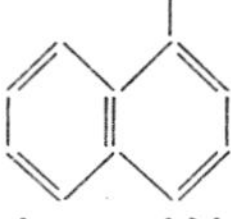

3. Aceto AN — Aceto
   Agerite Resin — Vanderbilt
   Antioxidant AP — Bayer
   Antivecchiante AL — ACNA Montecatini
   Nocrac C — Ouchi Shinko
4. Yellow-brown powder. M.W. 213.28; M.P. 145 °C. min.; S.G. 1.16.
   Staining oxi, ozo for NR, IR, BR, SBR, NBR.

**79** 1. Phenyl-alpha-naphthylamine

2.

3.

| | |
|---|---|
| Aceto PAN | Aceto |
| Antigene PA | Sumitomo |
| Antioxidant PAN | Bayer |
| Inibitore OA | ACNA Montecatini a |
| Neozone A | Du Pont |
| Nocrac PA | Ouchi Shinko |
| Nonox AN | ICI |

4. Light yellowish brown to violet lump. M.W. 219; M.P. 50 °C. min.; S.G. 1.21.
a. S.G. 1.16.
Staining oxi, flex, heat for NR, SBR, NBR, CR.

**80** 1. N,N'-Diaryl-p-phenylenediamine / phenyl-alpha-naphthylamine blend

2. —

3.

| | |
|---|---|
| Antigene FC | Sumitomo |

4. Dark-brown lump.
Staining oxi, heat, flex for NR, SBR, CR.

**81** 1. Phenyl-alpha-naphthylamine / 2,4-toluene diamine blend (92.5 : 7.5)

2. —

3.

| | |
|---|---|
| Neozone C | Du Pont |

4. Gray-brown pellets. M.P. 46 °C.; S.G. 1.22.
Staining oxi, heat for NR, SBR, IIR.

**82** 1. Phenyl-beta-naphthylamine

2.

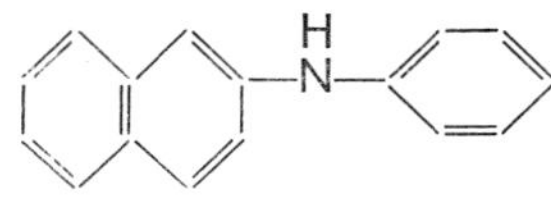


3.

| | |
|---|---|
| Accinox D | Alkali |
| Aceto PBN | Aceto |

| | |
|---|---|
| Agerite Powder | Vanderbilt |
| Antigene D | Sumitomo |
| Antigene DF | Sumitomo |
| Antioxidant 116 | Anchor |
| Antioxidant PBN | Bayer |
| Antioxidant PBN | Spolek |
| Inibitore OD | ACNA Montecatini |
| Neozone D | Du Pont |
| Neozone D Special (easy dispersable, for use in latex) | Du Pont |
| Nocrac D | Ouchi Shinko |
| Nonox D | ICI |
| Nonox D | ICI (India) |
| Nonox DN (powder and flakes) | ICI |
| PBN | BASF |
| PBN | Monsanto |
| PBN | Uniroyal |

4. Grayish-white powder. M.W. 219; M.P. 105 °C. min.; S.G. 1.23. Staining oxi, flex for NR, SBR, NBR, CR.

**83**

1. N,N'-Diaryl-p-phenylenediamine / phenyl-beta-naphthylamine blend
2. —
3. Akroflex DAZ (1 : 1) — Du Pont a
   Antigene HP-S — Sumitomo b
4. a. Dark-grey powder. M.P. 70 °C. min.; S.G. 1.27. Oxi, ozo for CR.
   b. Light-purple powder. M.P. 90 °C. min. Oxi, ozo, heat, flex for NR, SBR, NBR, CR.

**84**

1. N,N'-Diphenyl-p-phenylenediamine / phenyl-beta-naphthylamine blend
2. —
3. Agerite HP (65 : 35) — Vanderbilt
   Agerite HPX (Antioxidant 108X outside U.K.) — Anchor

| | |
|---|---|
| Akroflex CD (65 : 35) | Du Pont |
| Antigene HP | Sumitomo |
| Nocrac HP | Ouchi Shinko |
| Nonox HP | ICIANZ |

4. Grey to light-purple powder, rods, pellets. M.R. 89–94 °C.; S.G. 1.24.
   Staining oxi, flex for NR, SBR, NBR, CR.

**85**
1. N,N'-Diphenyl-p-phenylenediamine / 4,4'-dimethoxy-diphenylamine / N-phenyl-beta-naphthylamine blend (25 : 25 : 50)
2. —
3. Thermoflex A — Du Pont
4. Gray-to-black pellets. M.P. 78 °C.; S.G. 1.22.
   Oxi, ozo, flex, heat for NR, SBR, CR.

**86**
1. N,N'-Diphenyl-p-phenylenediamine / p-isopropoxy-diphenylamine / phenyl-beta-naphthylamine blend (25 : 25 : 50)
2. —
3. Agerite Hipar — Vanderbilt
4. Gray-to-brown powder and rods. M.R. 60–75 °C.; S.G. 1.16.
   Oxi.

**87**
1. Phenyl-beta-naphthylamine - acetone reaction product
2. —
3. Antigene DA — Sumitomo a
   Betanox Special — Uniroyal b
4. a. Light-brown powder. M.P. 65 °C. min.
   b. Tan powder. M.P. 130 °C. min.; S.G. 1.14.
   Oxi, heat, flex for NR, SBR, NBR.

**88**
1. N,N'-Diaryl-p-phenylenediamine / phenyl-beta-naphthylamine - acetone reaction product blend
2. —
3. Antigene FL — Sumitomo
4. Dark-gray powder.
   Heat, flex for NR, SBR, CR.

## Heterocyclic compounds

**89**
1. Benzofurane derivative
2. —
3. Antioxidant AFC — Bayer
4. Light-brown powder. M.P. 160 °C.; S.G. 1.25.
   Oxi, flex, heat for solid rubber and latex. Ozo for CR and for CR-mixtures.

**90**
1. 2-(3,5-Di-tert butyl-4-hydroxyanilino)-4,6-bis-(N-octylthio)-1,3,5-triazine
2. OH, $(CH_3)_3C$, $C(CH_3)_3$, NH, C, N, N, $H_{17}C_8$–S–C, C–S–$C_8H_{17}$, N
3. Irganox 565 — Ciba Geigy
4. White crystalline powder. M.W. 589; M.P. 91–96 °C.
   Oxi.

**91**
1. 6-Dodecyl-1,2-dihydro-2,2,4-trimethyl-quinoline
2. NH, $CH_3$, $CH_3$, $C_{12}H_{25}$, CH, $CH_3$
3. Santoflex DD — Monsanto
4. Dark viscous liquid. M.W. 341.56; S.G. 0.90–0.96.
   Staining oxi, flex for NR. FDA appr. Stabilizer for SBR during storage and production.

**92**

1. 6-Ethoxy-2,2,4-trimethyl-1,2-dihydroquinoline
2. [structural formula: $H_5C_2O$–, NH, $CH_3$, $CH_3$, CH, $CH_3$]
3. Trade names:

| Trade name | Manufacturer |
|---|---|
| Anox W | Bozzetto |
| Antioxidant EC | Bayer |
| Ethoxyquin | Bennett |
| Ethoxyquin | Raschig |
| Nocrac AW | Ouchi Shinko |
| Santoflex AW | Monsanto |

4. Dark red-brown viscous liquid. M.W. 217.30; Flash Point 162–164 °C.; S.G. 1.02–1.05.
   Staining oxi, ozo, flex for NR, SBR. FDA appr.

**93**

1. Ethylene-thiourea
   syn. 2-Mercaptoimidazoline
2. [structural formula: $H_2C$—N, $H_2C$, C–SH, N, H]
3. Trade names:

| Trade name | Manufacturer |
|---|---|
| Accelerator MI 12 | Metallgesellschaft |
| — | Degussa |
| E.T.U. | Hasselt |
| E.T.U. | Prochim |
| JOR 4022 | Bozzetto |
| JOR 4022 oleato (83 % JOR 4022, 17 % paraffin oil) | Bozzetto |
| NA–22 | Du Pont |
| NA–22 D (80 % dispersion of NA–22 in oil) | Du Pont |
| Nocceler 22 | Ouchi Shinko |
| Pennac CRA | Pennwalt |

| | |
|---|---|
| Pennac CRA | Vondelingenplaat |
| Robac 2.2 | Robinson |
| Rodanin S 62 | Dimitrova |
| Sanceler 22 | Sanshin |
| Soxinol 22 | Sumitomo |
| Vulkacit NPV/C (coated) | Bayer |

4. White powder. M.W. 102.16; M.P. 195 °C. min.; S.G. 1.43–1.45.
   Ozo for NR; non-staining acc. for CR.
   △ Acc. 27

**94** 1. 2-Mercaptobenzimidazole

2. [benzimidazole ring with N, C-SH, NH] ⇄ [benzimidazole ring with NH, C=S, NH]

3.

| | |
|---|---|
| Antigene MB | Sumitomo |
| Antioxidant MB | Dimitrova |
| Antioxidant MB | Bayer |
| Antivecchiante MB | ACNA Montecatini |
| MBI | Prochim |
| Nocrac MB | Ouchi Shinko |
| Permanax 21 | Rhône Poulenc |
| Vondantox MBI | Vondelingenplaat |

4. Yellow-white powder. M.W. 150; M.P. 290 °C. (with decomposition); S.G. 1.42.
   Non-staining oxi, heat, inh for NR, SBR, NBR; acc. for NR; pept. for CR.
   △ Acc. 86
   △ Pept. 3

**95** 1. Poly-2,2,4-trimethyl-1,2-dihydroquinoline

2. [ring structure: NH, $CH_3$, $CH_3$, CH, $CH_3$]$_n$

| | | |
|---|---|---|
| | 3. Aceto POD | Aceto |
| | Agerite AK (tradename outside U.K. Antioxidant 184) | Anchor |
| | Agerite Resin D | Vanderbilt |
| | Anox HB | Bozzetto |
| | Antigene RD | Sumitomo |
| | Antioxidant HS | Bayer a |
| | Flectol H | Monsanto |
| | Nocrac 224 | Ouchi Shinko |
| | Pennox HR | Pennwalt |
| | Permanax 45 | Rhône Poulenc |

4. Buff powder or flakes. M.P. 100–120 °C.; S.G. 1.08.
   a. M.P. 75 °C. min.
   Semi-staining oxi, heat, inh for NR, NBR, BR, IR, CR, also for latex; vulc. for CR.
   △ Vulc. 50

**96**

1. Tetraethyl-thiuram disulphide
2. $(C_2H_5)_2N.CS.S.S.CS.N(C_2H_5)_2$
3.

| | |
|---|---|
| Accicure TET | Alkali |
| Aceto TETD | Aceto |
| Ancazide ET | Anchor |
| Ethyl Thiram | Pennwalt |
| Ethyl Thiurad | Monsanto |
| Ethyl Tuads | Vanderbilt a |
| Ethyl Tuex | Naugatuck SpA |
| Ethyl Tuex | Uniroyal |
| Etiurac | Ticino |
| Eveite T | ACNA Montecatini |
| Hermat TET | Dimitrova |
| Nocceler TET | Ouchi Shinko |
| Robac TET | Robinson |
| Sanceler TET | Sanshin |
| Soxinol TET | Sumitomo |
| Superaccelerator 481 | Rhône Poulenc |
| TETD | Hasselt |
| TETD | Prochim |

| | |
|---|---|
| TETS | Bozzetto |
| Thiuram E | Du Pont |
| Vondac TET | Vondelingenplaat |
| Vulcafor TET | ICI |
| Vulcafor TET | ICI (India) |

4. Grayish-white powders and pellets. M.W. 296.54; M.P. 71–73 °C.; S.G. 1.26.
   a. Powder and rods.
   Non-staining stabilizer for Neoprene G; acc. for NR, SBR, NBR, IIR, IR, EPDM; act. for thiazole-, guanidine- and aldehyde-type acc.; vulc. for sulphurless compounds.
   △ Acc. 81
   △ Act. 8
   △ Vulc. 36

**97**
1. Trimethyldihydroquinoline-derivative
2. —
3. Antigene MW — Sumitomo
4. Dark-brown viscous liquid.
   Staining oxi, heat, flex for synthetic dry rubber and latex.

**98**
1. Zinc 2-mercaptobenzimidazole
2. 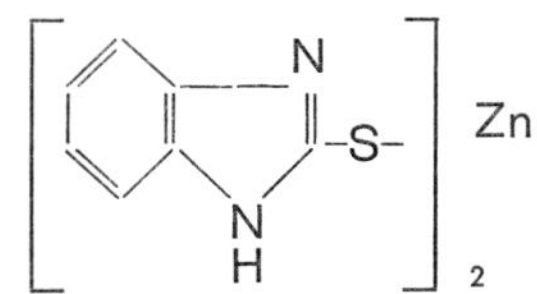

3. 

| | |
|---|---|
| Antigene MBZ | Sumitomo |
| Antioxidant ZMB | Bayer |
| M.B.I. Zn | Prochim |
| Nocrac MBZ | Ouchi Shinko |
| Permanax Z21 | Rhône Poulenc |

4. White powder. M.W. 363.4; M.P. 300 °C. (with decomposition); S.G. 1.6.
   Non-staining oxi, heat, inh for NR, SBR, NBR.

**99**
1. Zinc 2-mercaptobenzothiazole

2. (structural formula: zinc bis(2-benzothiazolyl thiolate), $Ar-S-Zn-S-Ar$)

3. | | |
|---|---|
| Accicure ZMBT | Alkali |
| Bantex | Monsanto |
| Eveite MZ | ACNA Montecatini |
| Hermat Zn MBT | Dimitrova |
| Nocceler MZ | Ouchi Shinko |
| OXAF | Naugatuck SpA |
| OXAF | Uniroyal |
| Pennac ZT | Pennwalt |
| Pennac ZT–"W" (containing 10 % inert hydrocarbon) | Pennwalt |
| Rapid Accelerator 205 | Rhône Poulenc |
| Sanceler MZ | Sanshin |
| Soxinol MZ | Sumitomo |
| Vulcafor ZMBT | ICI |
| Vulcafor ZMBT | ICI (India) |
| Vulkacit ZM | Bayer |
| Zenite | Du Pont |
| Zenite Special | Du Pont |
| Zetax | Vanderbilt |
| Zinc Ancap | Anchor |
| ZMBT | Cyanamid |
| ZMBT waxed | Cyanamid |
| ZMBT wettable | Cyanamid |

4. Light cream powder. M.W. 397.9; M.P. 300 °C. (with decomposition); S.G. 1.72.
   Non-staining oxi for latex; acc. for NR, SBR, NBR latices.
   △ Acc. 100

## Dithiocarbamates

**100** 1. Dibutylammonium-dibutyldithiocarbamate
2. $(C_4H_9)_2.N.CS.S.NH_2(C_4H_9)_2$
3. Robac D.B.U.D. — Robinson

4. Brown flakes. M.W. 334.6; M.P. 45 °C.
Oxi for rubber-based adhesives; non-staining acc. for NR- and SBR proofing compounds.
△ Acc. 46

**101** 1. Nickel dibutyldithiocarbamate
2. $[(C_4H_9)_2N.CS.S-]_2Ni$
3. | | |
|---|---|
| Antigene NBC | Sumitomo |
| JO 4011 | Bozzetto |
| NBC | Du Pont |
| NDBC | Hasselt |
| Nilame | Prochim |
| Nocrac NBC | Ouchi Shinko |

4. Green powder. M.W. 467.47; M.R. 86–90 °C.; S.G. 1.10.
Oxi, ozo, heat, flex for SBR, CR, CSM.

**102** 1. 2,2'-Dibenzothiazyl disulphide - morpholinium-N-oxy-diethylene-dithiocarbamate reacton product
2. —
3. | | |
|---|---|
| Antigene 3M | Sumitomo |

4. White powder. M.P. 105 °C.
Non-staining oxi, ozo, flex for NR, SBR.

**103** 1. Zinc dibutyldithiocarbamate
2. $[(C_4H_9)_2N.CS.S-]_2Zn$
3. | | |
|---|---|
| Aceto ZDBD | Aceto |
| Ancazate BU | Anchor |
| Butazate | Naugatuck SpA |
| Butazate | Uniroyal |
| Butazate 50D (50 % slurry for use in latex) | Uniroyal |
| Butazin | Ticino |
| Butyl Zimate | Vanderbilt |
| Butyl Ziram | Pennwalt |
| Butyl Ziram (50 % aquous solution) | Pennwalt |
| Butyl Zirame | Prochim |
| Eptac 4 | Du Pont |
| Eveite Butil Z | ACNA Montecatini |
| JO 4013 | Bozzetto |

| | |
|---|---|
| Nocceler BZ | Ouchi Shinko |
| Robac Z.B.U.D. | Robinson |
| Sanceler BZ | Sanshin |
| Soxinol BZ | Sumitomo |
| Ultra Accelerator DI 13 | Metallgesellschaft |
| Vondac ZBUD | Vondelingenplaat |
| Vulcafor ZNBC | ICI † |
| ZDBC | Hasselt |

4. Cream powder. M.W. 474.14; M.R. 95–108 °C.; S.G. 1.21.
   Antidegr. for unvulcanized rubber and for non-staining grades of IIR; non-staining ultra-acc. for NR, BR, SBR, NBR and their latices, acc. for EPDM; act. for thiazole- and other acid-type acc.
   △ Acc. 65
   △ Act. 28

**Miscellaneous, mixtures and undisclosed compositions**

**104** 1. Alkylated aryl-phosphites blend
2. —
3. Alkanox — Bozzetto
   Antioxidant 6 — Dimitrova
4. Amber liquid.
   Non-staining oxi for NR, SBR, BR.

**105** 1. 2,5-Di-tert amyl hydroquinone
2. OH; $CH_3$; $CH_3$; $CH_3CH_2$—C—; —C—$CH_2CH_3$; $CH_3$; $CH_3$; OH
3. Antioxidant DITAH — Conestoga
   Santovar A — Monsanto
   — — Eastman
4. Off-white powder. M.W. 250.37; M.P. 176 °C. min.; S.G. 1.02–1.08.
   Stabilizer for NBR; non-staining oxi for uncured rubber. FDA appr.

**106** 1. Diarylamine - ketone - aldehyde reaction product
2. —
3. BXA — Naugatuck SpA
BXA — Uniroyal
4. Brown powder. M.R. 85–95 °C.; S.G. 1.10.
Staining oxi, ozo, heat, flex for NR, SBR, NBR, CR.

**107** 1. complex Diarylamine - ketone reaction product / N,N'-diphenyl-p-phenylenediamine blend (65 : 35)
2. —
3. Flexamine — Uniroyal
Flexamine G — Uniroyal
4. Brown powder and granules. M.R. 75–90 °C.; S.G. 1.20.
Oxi, ozo, heat, flex, inh for NR, SBR, IR, BR.

**108** 1. 2,5-Di-tert butylquinol
syn. 2,5-Di-tert butyl-hydroquinone
2. OH
$-C(CH_3)_3$
$(CH_3)_3C-$
OH
3. Nocrac NS–7 — Ouchi Shinko
— — Eastman
— — May & Baker
4. Brownish-grey powder. M.W. 222.33; M.P. 200 °C.
Oxi for NR, IR, BR, SBR, NBR, CR.

**109** 1. N,N'-Dibutylthiourea
2. $(C_4H_9NH)_2CS$
3. Accelerator DBT — BASF
DBTU — Prochim
— — Degussa
Pennzone B — Pennwalt
Pennzone B — Vondelingenplaat
Robac DBTU — Robinson

4. Off-white powder. M.W. 188.3; M.P. 65 °C.
Antidegr. for NR-latex and for thermoplastic SBR; acc. for mercaptan-modified CR; act. for EPDM and NR.
△ Acc. 22
△ Act. 35

**110** 1. 1,3-Diethylthiourea
2. $(C_2H_5NH)_2CS$.
3. DETU — Prochim
JOR 4050 — Bozzetto
— — Degussa
Pennzone E — Pennwalt a
Pennzone E — Vondelingenplaat a
RC Granulat DETU (80 % DETU, 20 % saturated hydrocarbons and special dispersion agents) — Rhein Chemie
Robac DETU — Robinson
4. Yellow powder. M.W. 132.2; M.P. 75 °C.; S.G. 0.98. Water-soluble.
a. White flakes. S.G. 1.12.
Antidegr. for NR, NBR, SBR, CR; acc. for mercaptan-modified CR (Neoprene W).
△ Acc. 23

**111** 1. Dilaurylthiodipropionate
2. $(CH_2CH_2CO_2C_{12}H_{25})_2S$
3. Antigene TPL — Sumitomo
DLTDP — Reagens
Irganox PS 800 — Ciba Geigy
4. White crystalline powder or flakes. M.W. 514; M.P. 38 °C. min.
Oxi, ozo, heat, flex for NR, SBR.

**112** 1. p-Dimethoxybenzene
syn. Quinol-dimethylether
Hydroquinone-dimethylether

2. $OCH_3$ (benzene ring) $OCH_3$
3. — May & Baker
4. White crystalline solid. M.W. 138.2; M.P. 56 °C.; S.G. 1.053. Oxi.

**113** 1. N,N'-Diphenyl-p-phenylenediamine / acrylamide-derivative blend
2. —
3. Accinox DPL Alkali
4. Dark brown viscous liquid. S.G. 1.16. Oxi.

**114** 1. N,N'-Diphenyl-p-phenylenediamine / 6-dodecyl-1,2-dihydro-2,2,4-trimethylquinoline blend
2. —
3. Santoflex 75 Monsanto
4. Dark flakes. S.G. 1.09–1.15. Oxi.

**115** 1. Dioctadecylthiodipropionate
2. $(CH_2CH_2CO_2C_{18}H_{37})_2S$
3. Irganox PS 802 Ciba Geigy
4. White crystals. M.W. 682.0; M.P. 63–65 °C. Oxi, co-stabilizer.

**116** 1. Di-ortho-tolylguanidine salt of dicatecholborate
2. $CH_3$ ⊕ ⊖ —NH C=N H H —NH [ O O >B< O O ] $CH_3$

3. Nocceler PR — Ouchi Shinko
   Permalux — Du Pont
4. Grayish-brown powder. M.W. 482.8; M.P. 165 °C.; S.G. 1.25.
   Antidegr. for NR, SBR; non-staining acc. for Neoprene-G types.
   △ Acc. 33

**117** 1. Hydroquinone monobenzylether
2. $C_6H_5$—$CH_2$–O—$C_6H_4$—OH
3. Agerite Alba — Vanderbilt
4. Light-tan powder. M.W. 200; M.R. 108–115 °C.; S.G. 1.26.
   Oxi.

**118** 1. Hydroquinone-monomethylether
   syn. p-Methoxy phenol
2. $CH_3OC_6H_4OH$
3. MEHQ — Conestoga
4. Flakes. M.W. 124.13; M.P. 53 °C.
   Oxi.

**119** 1. 2-Mercaptobenzimidazole / sym. di-beta-naphthyl-p-phenylene-diamine blend
2. —
3. Nonox CGP — ICI †
4. Grey powder. S.G. 1.34.
   Oxi, inh.

**120** 1. 2-Mercaptobenzimidazole / 2,2'-methylene-bis-[6-(alpha-methyl-cyclohexyl)-p-cresol] blend
2. —
3. Nonox CNS — ICI
4. Cream powder. S.G. 1.25.
   Non-staining oxi, inh.

**121** 1. Nickel diisopropylxanthate
2. $(C_3H_7.O.CS.S–)_2Ni$
3. Sandant PN — Sanshin
4. Yellowish-green powder. M.W. 325.4; M.P. 110 °C.
   Flex.

**122** 1. Thiourea-derivatives
2. —
3. Nocrac NS–10–N — Ouchi Shinko
Nocrac NS–11 — Ouchi Shinko
4. Oxi for NR, IR, SBR, NBR, CR.

**123** 1. Tributylthiourea
2. $(C_4H_9)_2N.CS.NHC_4H_9$
3. Santowhite TBTU — Monsanto
4. Pale-amber liquid. M.W. 244; S.G. 0.938.
Non-staining ozo for natural and synthetic rubber.

**124** 1. Tris-(nonylated phenyl) phosphite
2. $\left[ C_9H_{19}—C_6H_4—O \right]_3$—P
3. Antigene TNP — Sumitomo
Anullex TNPP — Pearson
Irgafos TNPP — Ciba Geigy
Naugard — Uniroyal
(in Europe Polygard)
Grades P, P powder (68 % active ingredient on calcium-silicate carrier), PHR (hydrolysis resistant)
Nocrac TNP — Ouchi Shinko
Polygard — Rubber Regenerating
Polygard — Naugatuck SpA
4. Light-yellow tacky liquid. M.W. 689; S.G. 0.99.
Non-staining oxi, heat, flex for NR, SBR, NBR.

**125** 1. Composition undisclosed
2. —
3. Anox SD — Bozzetto
Antioxidans GI–08–288 — Ciba Geigy
Antioxidant 439 — Uniroyal
Antioxidant 451 — Uniroyal
Antioxidant TD EM 50 — Bayer
Antivecchiante ADD — ACNA Montecatini

| | |
|---|---|
| Arrconox DNP | Rubber Regenerating |
| Arrconox GP | Rubber Regenerating |
| Hallcolite OP | Hall |
| Inibitore AT | ACNA Montecatini |
| Kenmix | Kenrich a |
| Nevastain A | Neville |
| Nevastain B | Neville |
| Nonox | ICI |
| Grades CC, EX, EXN, EXP, NS, WSL, WSO | |
| Oxystop 999 | Arwal |
| Ozonschutzmittel AFD | Bayer |
| Ozonstop | Arwal |
| Grades 6000–E, 6000–P | |
| Raluquin K | Raschig |
| Wytox 345 | Nat. Polychemicals |

4. a. △ Acc. 141
△ Act. 44
△ Vulc. 54

# Blowing agents

## INORGANIC

**1**

1. Sodiumbicarbonate
2. $NaHCO_3$
3. — Anchor a
   Isocell S Rhein Chemie b
   Sponge Paste 2 Ouchi Shinko c
   Sponge Paste 3 Ouchi Shinko d
   Treibmittel 1843 Rhein Chemie e
   Unicel S Du Pont f
   Unicel SX Du Pont g
4. a. White powder. M.W. 84.02; M.P. 270 °C. (with decomposition); S.G. 2.15.
   b. Grayish coarse powder. Decomposition at 70 °C.; S.G. 2.15.
   c. 80 % $NaHCO_3$.
   d. 70 % $NaHCO_3$.
   e. Grayish coarse powder. Decomposition at 70 °C.; S.G. 2.15.
   f. 50 % Dispersion in mineral oil. Decomposition at 100 °C.; S.G. 1.30.
   g. 70 % Dispersion in mineral oil. Decomposition at 100 °C.; S.G. 1.55.

## ORGANIC

### Azo-compounds

**2**

1. Azo-dicarbonamide
   syn. 1,1-Azo-bis-formamide
2. $NH_2CON{:}NCONH_2$
3. Alveofer AZDC Bozzetto
   Azocel Fairmount
   Grades 504, 508, 525, varying in particle size.
   BZ M2 Organo Synthèse
   Celogen AZ Uniroyal

| | | |
|---|---|---|
| | Dispercel (50 % dispersion of Azocel in DOP) | Fairmount |
| | Genitron AC Grades AC/2, AC/3, AC/4 varying in particle size, also available as paste in plasticizer. (Trade name in N- and S-America and in Japan Ficel.) | Fisons |
| | Isocell AC | Rhein Chemie † |
| | Kempore | Nat. Polychemicals |
| | Noury ADC Grades ADC/4, ADC/8, ADC/25 varying in color and in particle size. | Noury |
| | Porofor ADC/R | Bayer |

4. Yellow powder. M.W. 116.08; S.G. 1.63; decomposition temperature in air 195–200 °C., in polymers 150–200 °C.

**3**

1. Azo-diisobutyronitrile
2. $(CH_3)_2.(CN).C–N=N–C(CN).(CH_3)_2$
3. Genitron AZDN (trade name in N- and S-America and in Japan Ficel) — Fisons
   Porofor N — Bayer
4. White powder. M.W. 164.2; S.G. 1.11; decomposition temperature 103 °C.

**4**

1. Diazoaminobenzene
2. $C_6H_5–N=N–N(H)–C_6H_5$
3. Vulcacel AN — ICI †
4. Brown powder. M.W. 197.

**Nitroso-compounds**

**5**

1. N,N'-Dimethyl-N,N'-dinitrosoterephthalamide

2. 

$$H_3C-\underset{\underset{NO}{|}}{N}-\overset{\overset{O}{\|}}{C}-C_6H_4-\overset{\overset{O}{\|}}{C}-\underset{\underset{NO}{|}}{N}-CH_3$$

3. Nitrosan — Du Pont
4. Yellow powder. M.W. 250; decomposition range 80–100 °C.; S.G. 1.20.
   Non-staining.

**6**

1. Dinitrosopentamethylenetetramine
2. 

$$\begin{array}{ccccc} & H_2C & — N — & CH_2 & \\ & | & | & | & \\ ON — & N & CH_2 & N & — NO \\ & | & | & | & \\ & H_2C & — N — & CH_2 & \end{array}$$

3. 

| | |
|---|---|
| Alveofer DNP 80 (80 % DNP, 20 % inert fillers) | Bozzetto |
| Alveofer DNP/GM (80 % DNP, 20 % inert fillers and dispersion agents) | Bozzetto |
| Chempor Grades B 80, N 90, PC 50, PC 65, PC 80. | Chemko |
| DNPT (74 % active ingredient) | Organo Synthèse |
| Esporen | ACNA Montecatini |
| Naftopor | Metallgesellschaft |
| Nocblow DPT | Ouchi Shinko |
| Opex Grades 40, 42, 80, 93, 100 (% DNPT) | Nat. Polychemicals |
| Porofor DNO/F (80 % DNPT, 20 % inert fillers) | Bayer |
| Porofor DNO/N (80 % DNPT, 20 % inert fillers) | Bayer |
| Sponge Paste 1 (50 % active ingredient) | Ouchi Shinko |
| Sponge Paste 4 (50 % active ingredient) | Ouchi Shinko |

Unicel — Du Pont
Grades 80, 93, 100 (% active ingredient) and ND (1 part Unicel 100 = 2.5 paste Unicel ND)
Vondablow — Vondelingenplaat
(80 % active ingredient)
Vulcacel BN 94 — ICI

4. Pale-yellow powder. M.W. 186.17; decomposition temperature in air 190–200 °C.; S.G. 1.51.

**Sulfo(nyl)hydrazide-compounds**

**7**

1. Benzene disulfohydrazide
2. $C_6H_4.(SO_2.NH\text{–}NH_2)_2$
3. Porofor B 13/CP 50 — Bayer
   (Paste with 50 % chlorinated paraffin)
4. Yellowish paste. M.W. 266; M.P. 147 °C.; S.G. 1.5; decomposition temperature 155 °C.

**8**

1. Benzene sulfohydrazide
2. $C_6H_5.SO_2.NH.NH_2$
3. BSH — Organo Synthèse
   Celogen BSH — Uniroyal
   Celogen BSH paste — Uniroyal
   (75 % BSH, 25 % paraffin oil)
   Genitron BSH — Fisons
   (trade name in N- and S-America and in Japan Ficel)
   Porofor BSH powder — Bayer
   Porofor BSH paste — Bayer
   (75 % BSH, 25 % paraffin oil)
   Porofor BSH paste M — Bayer
   (75 % BSH, 25 % paraffin oil, micronized)
4. Cream-to-buff powder. M.W. 172.2; decomposition temperature 99 °C.; S.G. 1.43.

**9**
1. p,p'-Oxy-bis (benzene sulphonyl) hydrazide
   syn. Diphenyloxide-4,4'-disulphonhydrazide
2. $NH_2.NH.SO_2.C_6H_4.O.C_6H_4.SO_2.NH.NH_2$
3. Celogen OT — Uniroyal a
   Genitron OB — Fisons b
   (trade name in N- and S-America and in Japan Ficel)
   Nitropore OBSH — Nat. Polychemicals c
4. White crystalline powder. M.W. 358.4
   a. S.G. 1.55; decomposition temperature 158–160 °C.
   b. S.G. 1.43; decomposition temperature 150 °C.
   c. S.G. 1.54; decomposition temperature 127–149 °C.

**10**
1. p-Toluene sulphohydrazide
2. $CH_3-C_6H_4-S(=O)(=O)-NH-NH_2$
3. Celogen TSH — Uniroyal
   Isocell TSH — Rhein Chemie †
4. Cream crystalline powder. M.W. 186; M.R. 100–110 °C.; S.G. 1.42; decomposition temperature 110–120 °C.
   For NR, SBR, CR and polysulphiderubber.

**11**
1. Sulphohydrazides
2. —
3. Alveofer BSI — Bozzetto
   Alveofer BSI Paste — Bozzetto
   (75 % BSI, 25 % paraffin oil)
4. White powder. M.P. 100 °C. min.; S.G. 1.4.

## Miscellaneous, mixtures and undisclosed compositions

**12**
1. Alkylsulphonate and activating additives blend
2. —
3. Levapon TH — Bayer
   (highly concentrated)

4. Yellowish paste. S.G. 1.04.
   Secondary blowing agent in combination with sodiumoleate to produce latex foams. Foam-stabilizer in the manufacture of solid cellular rubber goods.

**13**
1. Amino-guanidine bicarbonate
2. $[NH_2.NH.C(:NH).NH_2].HCO_3$
3. Genitron OC — Fisons
   (trade name in N- and S-America and in Japan Ficel)
4. Off-white powder. M.W. 136.1. Decomposition temperature 167 °C.

**14**
1. p-Toluene-sulfonyl-semicarbazide
2. $CH_3.C_6H_4.SO_2.NH.NH.CO.NH_2$
3. Celogen RA — Uniroyal
4. Off-white powder. M.W. 136.1; decomposition temperature 167 °C.

**15**
1. Zinc-amine complex
2. —
3. Ancablo A — Anchor
4. White powder.
   Ba. for NR, SBR, IIR, CR.

**16**
1. Composition undisclosed
2. —
3. Genitron CR — Fisons a
   Genitron THT — Fisons b
   (trade name in N- and S-America and in Japan Ficel)
   OP–2 — Organe Synthèse c
4. a. Yellow powder. Decomposition temperature 150–200 °C.
   b. Off-white powder. Decomposition temperature 265–290 °C.
   c. Yellow powder. Decomposition temperature 150–160 °C.

# Co-agents

**1**

1. 1,3-Butylideneglycoldimethacrylate
2. $CH_2{=}C(CH_3){-}C({=}O){-}O{-}CH_2{-}CH_2{-}CH(CH_3){-}O{-}C({=}O){-}C(CH_3){=}CH_2$
3. SR 297 — Anchor
   SR 297 — Sartomer
4. Clear straw coloured liquid. M.W. 226; B.P. 290 °C.; S.G. 1.009.
   Co-agent for the peroxide vulc. of EPDM and NBR.

**2**

1. Ethylenedimethacrylate
   syn. Ethyleneglycoldimethacrylate
2. $CH_2{=}C(CH_3){-}C({=}O){-}O{-}CH_2{-}CH_2{-}O{-}C({=}O){-}C(CH_3){=}CH_2$
3. SR 206 — Anchor
   SR 206 — Sartomer
4. Clear, water-white liquid. M.W. 198; B.P. 260 °C.; S.G. 1.05.
   Co-agent in peroxide vulcanization of EPDM and NBR.

**3**

1. Phenolformaldehyde resin
2. —
3. Vulkadur A — Bayer
4. White to yellow powder. M.P. 75–90 °C.; S.G. 1.27.
   Reinforcing resin for NBR.

**4**

1. N,N'-m-Phenylene-dimaleimide
2. Structure:

```
      O                 O
      ‖                 ‖
   HC–C                 C–CH
   ‖    >N–(C6H4, m)–N<   ‖
   HC–C                 C–CH
      ‖                 ‖
      O                 O
```

3. HVA 2 — Du Pont
4. Yellow powder. M.W. 268; M.P. 200 °C.; S.G. 1.44.
   Co-agent especially for CSM vulcanization.

**5**

1. Trimethylolpropanetrimethacrylate
2.

```
                    CH3
                    |
           CH2 = C - C - O
                     ||  |
                     O   |
       CH3               CH2
       |                 |
CH2 = C - C - O - CH2 - C - CH2 - CH3
          ||             |
          O              |
                   CH3   CH2
                   |     |
            CH2 = C - C - O
                      ||
                      O
```

3. SR 350 — Anchor
   SR 350 — Sartomer
4. Clear straw-coloured liquid. M.W. 338; B.P. 200 °C. at 1 mm min.

   Co-agent for the peroxide vulcanization of NBR and EPDM.

# Peptizers

---

**1**

1. 4-tert Butyl-o-thiocresol
2. (structural formula: $CH_3$–C($CH_3$)($CH_3$)–C$_6$H$_3$($CH_3$)–SH)
3. Pitt Consol 646 — Pitt Consol
   Pitt Consol 646 powder (55 % active ingredient on $SiO_2$ filler) — Pitt Consol
4. 55 % Hydrocarbon solution; colourless, low-viscosity liquid. Flash-Point 68.3 °C.; S.G. 0.87–0.90.
   Pept. for NR, SBR and other synthetic rubbers.

**2**

1. o,o'-Dibenzamido-diphenyl disulphide
   syn. Di-thio-bis-benzanilide
   Di-(o-benzamidophenyl) disulphide
2. (structural formula: C$_6$H$_5$–CO–NH–C$_6$H$_4$–S–S–C$_6$H$_4$–NH–CO–C$_6$H$_5$, printed as —NH—CO— … —S—S— … —CO—NH—)
3. Noctizer SS — Ouchi Shinko
   Peptazin BAFD — Dimitrova
   Peptisant 10 — Rhône Poulenc
   Pepton 22 — Cyanamid
   Pepton 22 — Anchor
4. Pale-yellow powder. M.W. 456.59; M.P. 136 °C.; S.G. 1.35.
   Pept. for NR, BR, IR, SBR.

**3**

1. 2-Mercaptobenzimidazole
2. (structural formula: benzimidazole with N, NH, C–SH ⇄ benzimidazoline with NH, NH, C=S)
3. Antigene MB — Sumitomo
   Antioxidant MB — Dimitrova
   Antioxidant MB — Bayer

| | |
|---|---|
| Antivecchiante MB | ACNA Montecatini |
| MBI | Prochim |
| Nocrac MB | Ouchi Shinko |
| Permanax 21 | Rhône Poulenc |
| Vondantox MBI | Vondelingenplaat |

4. Yellow-white powder. M.W. 150; M.P. 290 °C. (with decomposition); S.G. 1.42.

Pept. for CR; acc. for NR; non-staining oxi, heat, inh for NR, SBR, NBR.

△ Acc. 86
△ Antidegr. 94

**4**

1. 2-Mercaptobenzothiazole
2. 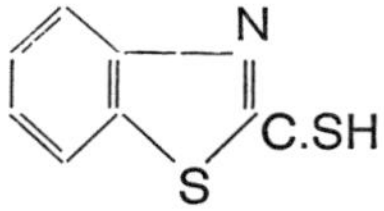 (benzene ring fused to thiazole ring: N, S, C.SH)
3. Trade names:

| | |
|---|---|
| Accicure MBT | Alkali |
| Ancap | Anchor |
| Captax | Vanderbilt |
| Eveite M | ACNA Montecatini |
| MBT | Akron |
| MBT | Du Pont |
| MBT | Naugatuck SpA |
| MBT | Uniroyal |
| MBT | Cyanamid |
| MBT XXX (very pure grade) | Cyanamid |
| Nocceler M | Ouchi Shinko |
| Pennac MBT | Pennwalt |
| Pneumax MBT | Dimitrova |
| Rapid Accelerator 200 | Rhône Poulenc |
| Rotax | Vanderbilt |
| Sanceler M | Sanshin |
| Soxinol M | Sumitomo |
| Thiotax MBT | Monsanto |
| Vulcafor MBT | ICI |
| Vulcafor MBT | ICI (India) |
| Vulkacit Merkapto | Bayer |

4. Light yellow powder. M.W. 167.25; M.R. 164–175 °C.; S.G. 1.50.
   Pept. for NR; non-staining semi-ultra acc. for NR, SBR, NBR, IIR, also for latex; ret. for CR.
   △ Acc. 96
   △ Ret. 6

**5**
1. Pentachlorothiophenol (with additives)
   syn. Pentachlorophenyl-mercaptan
2. $C_6Cl_5SH$
3. Ciclizante 2°B — Bozzetto a
   Ciclizante 3°C — Bozzetto b
   Renacit V — Bayer c
   Renacit VII — Bayer d
   RPA No. 6 — Du Pont e
4. Gray powder. M.W. 280; S.G. a. 1.15, b. 1.17, c. 1.68, d. 2,33, e. 1.83.
   Pept. for NR, SBR, NBR, BR, IIR.

**6**
1. Piperidinium-pentamethylenedithiocarbamate
   syn. N-Pentamethyleneammonium-N-pentamethylenedithiocarbamate
2. $(CH_2)_5N.CS.S.H_2N(CH_2)_5$
3. Accelerator 552 — Du Pont
   Accelerator 2P — Anchor
   Nocceler PPD — Ouchi Shinko
   Pentalidine — Prochim
   Robac P.P.D. — Robinson
   Vulkacit P — Bayer
4. Cream powder. M.W. 246; M.P. 175 °C.; S.G. 1.19.
   Pept. for Neoprene G and KNR types; acc. for NR, NBR, SBR; act. for thiuram- and thiazole-type acc.
   △ Acc. 54
   △ Act. 22

**7**
1. Thioxylenol
   syn. Xylenethiol
   Xylyl mercaptan
2. $C_6H_3(CH_3)_2SH$

3. Pitt Consol 640 — Pitt Consol a
   RPA No. 3 — Du Pont b
   RPA No. 3 conc. — Du Pont c
4. a. 40 % Solution in hydrocarbon solvent. M.W. 138; Flash Point 68.3 °C; S.G. 0.90–0.93.
   b. Amber liquid, (diluent petroleum oil). M.W. 139; Flash Point 74 °C.; S.G. 0.90.
   c. Concentration of RPA No. 3 conc. = 2 × RPA No. 3. Flash-Point 82 °C.; S.G. 1.00.
   Pept. for NR, SBR, IR.

**8**

1. Zinc 2-benzamidothiophenate
2. HN—OC—C₆H₅ ... C₆H₅—CO—NH
   (bis-[2-(benzoylamino)phenyl]: —S—Zn—S—)
3. Noctizer SZ — Ouchi Shinko
   Pepton 65 — Cyanamid
   Pepton 65 — Anchor
4. Off-white powder. M.W. 521.96; M.R. 200–230 °C.; S.G. 1.32.
   Pept. for NR, SBR, NBR, IR.

**9**

1. Zinc benzoyldisulphide
   syn. Zinc. thiobenzoate
2. $(C_6H_5COS)_2Zn$
3. Peptizer No. 2 — Sanshin a
   Robac TBZ — Robinson b
4. Off-white powder.
   a. M.W. 339.75; M.P. 110 °C. min.; S.G. 1.45–1.50.
   b. Hydrate (2 $H_2O$). M.W. 375.8; M.P. 110 °C.
   Non-staining pept. for NR, SBR, IR.

**10**

1. Zinc tert butylthiophenate
2. $(CH_3)_3C—C_6H_4—S—Zn—S—C_6H_4—C(CH_3)_3$

3. Pitt Consol 651 — Pitt Consol a
   Pitt Consol 651 conc. — Pitt Consol b
   Pitt Consol 651 A — Pitt Consol c
4. White powder, extended with inert fillers. M.W. 395; S.G. a. 1.41, b. 1.22, c. 1.80.
   Pept. for NR, SBR and other synthetic rubbers.

**11**
1. Zinc oxide dispersions
2. ZnO
3. Polyzinc — Polychimie
   Grades A, A75, V–01–70
4. Predispersed ZnO.
   Pept. for NR, SBR, NBR, CR and polyblends.

**12**
1. Zinc salt of pentachlorothiophenol
2. $(C_6Cl_5S-)_2Zn$
3. Ciclizante 1°A — Bozzetto
   Endor — Du Pont
   Renacit IV — Bayer
   Renacit IV/GR — Bayer a
   (Renacit IV 66 %, stearic acid and paraffin 34 %)
4. Blueish-gray powder. M.W. 623; M.P. 335 °C. (with decomposition); S.G. 2.33.
   a. Gray granules. M.P. 50–55 °C.; S.G. 1.53.
   Pept. for NR, IIR and SBR (with oil).

**13**
1. N,S-substituted derivative of o-Aminothiophenol
2. —
3. Peptazin X/BFT — Dimitrova
4. White powder. M.P. 145 °C. min.
   Pept. for NR.

**14**
1. 2-Naphthalene-thiol / paraffinwax blend (33 : 67)
2. —
3. RPA No. 2 — Du Pont
   Premax 2 — ACNA Montecatini
4. Cream coloured flakes. M.W. 160.22; Flash Point 160 °C.; S.G. 0.92.
   Pept. for NR, SBR.

**15**
1. Sulphonated petroleum products blend
2. —
3. Ancoplas — Anchor
   Grades OB, ER, LP.
4. Pept. for NR, SBR and reclaim rubber.

**16**
1. o,o'-Dibenzamido-diphenyl disulphide / zinc 2-benzamidothiophenate blend
2. —
3. Noctizer SM — Ouchi Shinko
4. M.P. 130 °C.
   Pept. for NR, IR, BR, SBR.

**17**
1. Thio-beta-naphthol / inert wax blend (33 : 67)
2. —
3. Vulcamel TBN — ICI †
4. White waxy flakes. M.P. 50 °C.; S.G. 0.9.

**18**
1. Zinc 2-benzamidothiophenate / zinc thiobenzoate blend
2. —
3. Noctizer SX — Ouchi Shinko
4. M.P. 103 °C.
   Pept. for NR, IR, BR, SBR.

**19**
1. Composition undisclosed
2. —
3. Aktiplast — Rhein Chemie a
   Aktiplast T — Rhein Chemie a
   Bondogen — Vanderbilt b
   Dispergum N — Grandel c
   Struktol A 50 — Schill & Seilacher a
4. a. Yellow powder and pellets. M.P. ca. 80 °C.; S.G. 1.1.
      Pept. for NR and IR, not usable in CR compounds.
   b. Dark brown liquid. S.G. 0.91–0.93.
      General purpose.
   c. Yellow paste. M.R. 70–75 °C.
      Not usable in CR compounds.

# Retarders

---

**1**

1. Benzoic acid
2. $C_6H_5COOH$
3. Benzoic acid GK — Bayer
   Benzoic acid GV — Bayer
   Retarder BA — Akron
   Ritardante AB — ACNA Montecatini a
4. White crystalline powder. M.W. 122.12; M.P. 122 °C.; S.G. 1.26.
   Non-staining retarder for NR, SBR, NBR, and their latices.
   a. Benzoic acid with a dispersion agent. M.P. 112–118 °C.; S.G. 1.20.

**2**

1. 4,4'-Diamino diphenyl methane
2. $CH_2(C_6H_4.NH_2)_2$
3. Robac 4.4 — Robinson a
   Tonox — Uniroyal b
4. a. Light brown powder. M.W. 198.3; M.R. 75–85 °C.
   Ret. for IIR; acc. for CR; anti-frosting agent for NR.
   b. Brown waxy lump. S.G. 1.18.
   △ Acc. 15
   △ Antidegr. 42

**3**

1. 2,2'-Dibenzothiazyl disulphide
   syn. Di-(2-benzothiazyl) disulphide
2. [Structural formula: two benzothiazole rings (N, S) linked at C-2 by –S–S–]
3. Accicure MBTS — Alkali
   Altax — Vanderbilt
   Ancatax — Anchor
   Bowax AC/MBTS — Bozzetto
   (MBTS dispersed in Bowax C)
   Eveite DM — ACNA Montecatini
   MBTS — Akron
   MBTS — Cyanamid
   MBTS — Du Pont

| | |
|---|---|
| MBTS | Naugatuck SpA |
| MBTS | Uniroyal |
| Mercasulf MBTS | Bozzetto |
| Nocceler DM | Ouchi Shinko |
| Pennac MBTS | Pennwalt |
| Pneumax DM | Dimitrova |
| Pneumax F | Dimitrova a |
| Rapid Accelerator 201 | Rhône Poulenc |
| Sanceler DM | Sanshin |
| Soxinol DM | Sumitomo |
| Thiofide MBTS | Monsanto |
| Vulcafor MBTS | ICI |
| Vulcafor MBTS | ICI (India) |
| Vulkacit DM | Bayer |

4. Yellowish powder. M.W. 332.50; M.P. 170–175 °C.; S.G. 1.50.
   a. mixture with basic acc.
   Non-staining. Ret. for CR; semi-ultra acc. for NR, NBR, IIR, SBR.
   △ Acc. 92

**4**

1. Dibutylammoniumoleate
2. 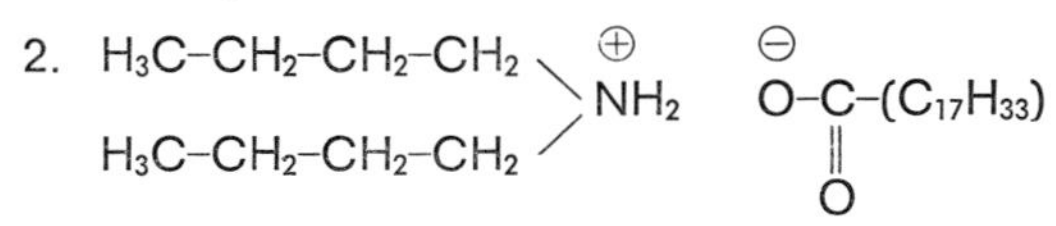

3. | | |
   |---|---|
   | Activator 1102 | Anchor |
   | Barak | Du Pont |
   | DOB | Organo Synthèse |
4. Dark amber liquid. M.W. 409; Flash Point 102 °C.; S.G. 0.88.
   Act.-ret. for thiuram-type acc.; act. for thiazole-, thiuram- and sulphenamide-type acc.
   △ Act. 34

**5**

1. Dicyandiamine / phtalic anhydride blend
2. —
3. Retarder AK — Conestoga
4. White powder. M.P. 123–132 °C.; S.G. 1.48.
   Non-staining ret. for NR, NBR, SBR.

**6**

1. 2-Mercaptobenzothiazole
2. N
   C.SH
   S
3. Trade names:

| Trade name | Manufacturer |
|---|---|
| Accicure MBT | Alkali |
| Ancap | Anchor |
| Captax | Vanderbilt |
| Eveite M | ACNA Montecatini |
| MBT | Akron |
| MBT | Cyanamid |
| MBT XXX (very pure grade) | Cyanamid |
| MBT | Du Pont |
| MBT | Naugatuck SpA |
| MBT | Uniroyal |
| Nocceler M | Ouchi Shinko |
| Pennac MBT | Pennwalt |
| Pneumax MBT | Dimitrova |
| Rapid Accelerator 200 | Rhône Poulenc |
| Rotax | Vanderbilt |
| Sanceler M | Sanshin |
| Soxinol M | Sumitomo |
| Thiotax (MBT) | Monsanto |
| Vulcafor MBT | ICI |
| Vulcafor MBT | ICI (India) |
| Vulkacit Merkapto | Bayer |

4. Light yellow powder. M.W. 167.25; M.R. 164–175 °C.; S.G. 1.50. Ret. for CR; non-staining acc. for NR, SBR, NBR, IIR, also for latex; pept. for NR.
   △ Acc. 96
   △ Pept. 4

**7**

1. N-Nitroso-diphenylamine
2. 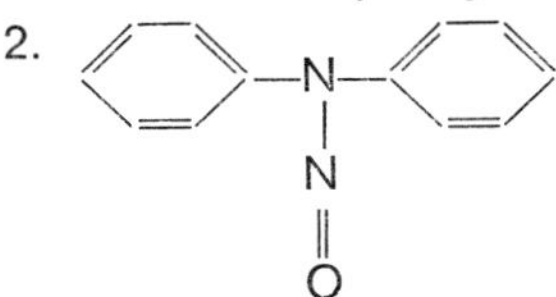

3. Accitard A — Alkali
   Curetard A — Monsanto
   Goodrite Vultrol — Goodrich
   N.D.A. — Prochim
   Redax — Vanderbilt
   Retarder J — Uniroyal
   Retarder 2N — Cyanamid
   Retarder NDPA — Conestoga
   Retrocure — Akron
   Ritardante AN — ACNA Montecatini
   Sconoc — Ouchi Shinko
   Vulcatard A — ICI
   Vulcatard A — ICI (India)
   Vulkalent A — Bayer
   Wiltrol N — Nat. Polychemicals
4. Brown granules. M.W. 198; M.P. 65 °C.; S.G. 1.27.
   Staining ret. for NR, NBR, SBR.

**8**

1. Phthalic anhydride
2. $C_6H_5(CO)_2O$
3. ESEN — Uniroyal a
   PA — Raschig
   PSA — Raschig
   Retarder AK — Akron
   Retarder B–C — Sanshin a
   Retarder PD — Cyanamid b
   Retarder PD — Anchor b
   Ritardante AF — ACNA Montecatini
   Sconoc 5 (70 % active) — Ouchi Shinko
   Sconoc 7 — Ouchi Shinko
   Sumitard BC — Sumitomo
   Vulkalent B/C (surface coated) — Bayer a
   Wiltrol P — Nat. Polychemicals a
4. White powder. M.W. 148.12; M.P. 130 °C. min.; S.G. 1.51.
   a. Surface treated.
   b. Modified.
   Non-staining ret. for NR, SBR, IIR.
   Non-usable in CR-compounds.

**9**

1. Salicylic acid
   syn. ortho-Hydrobenzoic acid
2. $HOC_6H_4OOH$
3. Retarder W — Du Pont
   Retarder TSA — Monsanto a
   Ritardante AS — ACNA Montecatini
   Salicylic R — Monsanto
4. Buff powder. M.W. 138.12; M.P. 159 °C.; S.G. 1.37.
   a. White powder. M.P. 155 °C.; S.G. 1.40., with dispersion agents.

**10**

1. Tetrabutyl-thiuram disulphide
2. $[(C_4H_9)_2N.CS.S-]_2$
3. Robac T.B.U.T. — Robinson
   Soxinol TBT — Sumitomo
   TBTS — Bozzetto
4. Brown liquid. M.W. 408.7; solidifies at ca. 20 °C.; S.G. 1.1. Insoluble in water.
   Ret. for CR. Non-staining acc. in combination with P.T.D. or T.M.T. for NR, SBR, NBR in sulphurless compounds; vulc. for NR, SBR, NBR.
   △ Acc. 79
   △ Vulc. 35

# Vulcanizing agents

## INORGANIC

**1**

1. Lead oxides
2. —
3. Mix Lpb 80 — Bozzetto a
   Mix Pb 80 — Bozzetto b
   Polyminium — Polychimie b
   Polytharge — Polychimie a
   grades A, B.D
   RC Granulat PbO — Rhein Chemie a
   RC Granulat $Pb_3O_4$ — Rhein Chemie b
   — — Anchor c
4. a. PbO dispersion. Acc. for CR; act. for NR, NBR, SBR, IIR; vulc. for CR, CSM.
   b. $Pb_3O_4$ dispersion. Acc. for CR; act. for IIR.
   c. PbO powder (litharge). Vulc. for CR.
   △ Acc. 1
   △ Act. 1

**2**

1. Magnesium oxide
2. MgO
3. Kenmag — Kenrich a
   Maglite — Merck b
   grades D, L, K, M, Y
   RC Granulat MgO (80 % MgO, 20 % saturated hydrocarbons and dispersion agents) — Rhein Chemie c
   Scorchguard — Anchor d
   grades C3, O, W
   Scorchguard O — Newalls e
   Struktol — Schill & Seilacher f
   grades WB 900, WB 902 (coated)
4. M.W. 40.32.
   a. S.G. 2.02. Act. for CR.

b. White powders. S.G. 3.3–3.5.
Vulc. for CR, CSM; act. for SBR and fluoro-elastomers; antidegr. for CR, CSM, chlorobutyl, fluoro-elastomers, SBR.
c. S.G. 2.06.
d. Grade C3, powder, heavy calcined MgO. Grade O, putty, light calcined MgO. Grade W, powder, light calcined MgO.
e. Dispersion. S.G. 2.08.
f. Act. for CR and CSM mixtures.
△ Act. 2
△ Antidegr. 1

**3**
1. Selenium
2. Se
3. Ancasal — Anchor
Vandex — Vanderbilt
4. Dark-gray powder. M.P. 217 °C. min.; S.G. 4.80.
Sec. vulc.

**4**
1. Sulphur
2. S
3. Crystex — Stauffer
Grades 90, OT (insoluble sulphur); tire Brand grades 21–1, 21–4, 21–7, 21–10; conditioned tire Brand grades 21–10TP, 21–12MC, 21–13, 21–14, 52–AF.
Insoluble Sulphur 60 — Monsanto
Manox Brand — Anchor
Grades insoluble sulphur, oiled insoluble sulphur, MC sulphur ($MgCO_3$ coated), 9;1 oil treated sulphur.
Polysoufre — Polychimie
Grades SE–01–70, SV–01–80, SN–05–75, SA–01–75, IV–01–75, IV–01–43.

RC Granulat S — Rhein Chemie
(80% sulphur, 20% saturated hydrocarbons and dispersing agents)
RC Schwefel Extra — Rhein Chemie
(93 % sulphur, 7 % dispersing agents)
S 84 — Bozzetto
(predispersed)
Colloidal Sulphur — Bayer
(95 %)
Struktol — Schill & Seilacher
Grades SU 95, (coated, soluble), SU 105, (paste, soluble), SU 106, (coated, insoluble), SU 108, (coated, insoluble), SU 120, (coated, soluble), SU 135, (oil treated, insoluble).

4. —

**5**

1. Sulphur dichloride
2. $SCl_2$
3. — Bayer
4. Yellow-to-reddish-yellow liquid. M.W. 102; B.R. 133–141 °C.; S.G. 1.68.
   For cold vulcanization of thin-walled rubber goods.

**6**

1. Tellurium
2. Te
3. Ancatel — Anchor
   Telloy — Vanderbilt
4. Grey powder. M.P. 450 °C.; S.G. 6.26.
   Sec. vulc.

**7**

1. Basic zinccarbonate
2. $ZnCO_3.2\ ZnO.3\ H_2O$

3. Zinkoxid Transparent — Bayer
   — — Durham
4. White powder. S.G. 3.3–3.5.
   Vulc. for CR; act. for sulphur- and peroxide vulcanization of NR, IR, BR, SBR, NBR, IIR, CR, EPDM, CSM.
   △ Act. 5

## ORGANIC

### Phenols

**8**

1. Alkylphenol disulphide
2. —
3. Vultac — Pennwalt
   Grades 2, 3, 4, 5, 6.
   Vultac — Vondelingenplaat
   Grades 2, 3, 5.
4. Grade 2: dark-brown tacky solid. M.R. 50–60 °C.; S.G. 1.1–1.2.
   Grade 3: dark-brown tack-free solid. M.R. 78–93 °C.; S.G. 1.15–1.25.
   Grade 4: stearic acid blend. Tacky brown solid. M.R. 48–58 °C.; S.G. 1.05–1.15.
   Grade 5: brown powder. S.G. 1.435.
   Grade 6: amber-brown liquid. S.G. 1.04–1.09.
   Vulc. for NR, SBR, NBR, chlorobutyl- and bromobutyl rubbers.

**9**

1. Alkylphenol resins
2. —
3. Vulkaresat — Reichhold Albert a
   Grades 510 E, 532 E.
   Vulkaresen — Reichhold Albert b
   Grades E 71, 105 E, 130 E.
   Vulkaresen — American Hoechst
   Grades PA 105, PA 130, PA 510.
4. a. Grade 510 E: M.R. 60–65 °C. Vulc. for EPT.
   Grade 532 E: M.R. 60–65 °C. Vulc. for EPT.

b. Grade E 71: M.R. 60–75 °C. Vulc. for IIR, halogenated IIR, NBR.
Grade 105 E: M.R. 54–60 °C. Vulc. for NR, SBR.
Grade 130 E: M.R. 60–65 °C. Vulc. for NR, SBR, NBR, IIR.

**10**
1. Bisphenol derivatives
2. —
3. Vulkaresol 315 E — Reichhold Albert
4. 70 % solution in a mixture of butanol and xylene.
   Vulc. for all types of rubber.

**11**
1. Octyl-phenol resin
2. —
3. Varcum 29–530(1198) — Reichhold
4. Light yellow powder. M.R. 80–90 °C.; S.G. 1.03–1.05.
   Vulc. for IIR.

**12**
1. Phenolformaldehyde resin (reactive)
2. —
3. Arrcorez 16 — Uniroyal
4. Yellow-to-brown brittle resin in lump form. M.P. 60–70 °C.; S.G. 1.05.
   Vulc. for IIR. FDA appr.

## Amines

**13**
1. Cumenediamine / m-phenylenediamine blend (58 : 42)
2. —
3. Caytur 7 — Du Pont
4. Dark-amber-to-black liquid. Freezing Point 16 °C.; S.G. 1.08.
   Curing agent for liquid urethane elastomers.

**14**
1. N,N'-Di-cinnamylidene-1,6-hexanediamine
   syn. N,N'-Di-cinnamylidene-1,6-hexamethylenediamine
2. $C_6H_5-CH=CH-CH=N-(CH_2)_6-N=CH-CH=CH-C_6H_5$
3. Diak No. 3 — Du Pont a
   Tecnocin A — ACNA Montecatini b

4. Tan coarse powder. M.W. 344; M.R. 82–88 °C.; a. S.G. 1.09, b. S.G. 0.374.
   Vulc. for fluoroelastomers.

**15** 1. 4,4'-Methylene-bis-(2-chloro-aniline)
   syn. 4,4'-Methylene-bis-(o-chloro-aniline)
2. $CH_2(C_6H_3NH_2Cl)_2$
3. Cyanaset M — Cyanamid
   MOCA — Du Pont
4. Light tan pellets. M.W. 267; M.R. 100–109 °C.; S.G. 1.44.
   Curing agent for urethane elastomers.

**16** 1. Dichlorobenzidine / methylene-bis-(o-chloroaniline) blend
2. —
3. Cyanaset H — Cyanamid a
   Cyanaset S — Cyanamid b
4. Large-grain dark-gray powder. S.G. 1.44; a. M.R. 106–110 °C.; b. M.R. 92–100 °C.
   Curing agent for urethane elastomers.

**Peroxides**

**17** 1. Benzoyl peroxide
2. $C_6H_5\text{–}C(=O)\text{–O–O–}C(=O)\text{–}C_6H_5$
3. Cadox — Nourychem a
   Grades BSD, BSG.
   Lucidol S 50 — Noury b
   Norox BZP–S–50 — Norac c
4. M.W. 242
   a. Grade BSD: 50 % silicone-oil paste (D.C. 200). S.G. 1.12.
      Grade BSG: 50 % silicone-oil paste (Gen. Electric SF–96–1000). S.G. 1.12.
   b. 50 % silicone-oil paste.
   c. 50 % paste in polydimethylsiloxane 1000.
   Vulc. for siliconerubber.

**18**

1. 1,3-Bis-(tert butyl-peroxy-isopropyl) benzene
2. $CH_3-C(CH_3)_2-O-O-C(CH_3)_2-C_6H_4-C(CH_3)_2-O-O-C(CH_3)_2-CH_3$
3. Perkadox — Noury a
   Grades 14, 14/40
   Peroximon F 40 — ACNA Montecatini b
   Retilox F 40 — ACNA Montecatini b
4. a. Grade 14; pale-yellow crystalline powder. M.W. 338; M.P. ca. 50 °C.; S.G. 1.1.
   Grade 14/40, 40 % active ingredient, 60 % $CaCO_3$. Granular. S.G. 1.16.
   b. With inert filler. White small cylinders. M.P. 35 °C.; S.G. 1.43.
   Vulc. and crosslinking agent for natural- and synthetic rubbers, siliconerubber and polyolefins.

**19**

1. Butyl-4,4-bis-(tert butyl peroxy) valerate
2. $H_3C-C(O-O-C(CH_3)_3)_2-(CH_2)_2-C(=O)-O(CH_2)_3CH_3$
3. Luperco 230 XL — Lucidol a
   Lupersol 230 — Lucidol b
4. a. White powder, 50 % active ingredient on an inert filler. Bulk dens. 24 lbs/cu ft.
   b. Colourless liquid. M.W. 334.46; S.G. 0.9503.
   Curing agent for EP rubber, EPT, SBR, silicone rubber and urethane elastomers.

**20** 1. p-Chlorobenzoyl peroxide

2. $Cl-C_6H_4-C(=O)-O-O-C(=O)-C_6H_4-Cl$

3. Cadox PS — Nourychem

4. 50 % Silicone-oil paste, containing 10 % dibutylphthalate. S.G. 1.17.

Curing agent for silicone rubber.

**21** 1. Cumene hydroperoxide

2. $C_6H_5-C(CH_3)_2-O-OH$

3. Trigonox K 70 — Noury

4. Colourless liquid, 70% active ingredient in a mixture of alcohols, ketones and cumene. Flash Point 61 °C.; S.G. 1.01–1.04.

Curing agent for polysulphide rubber.

**22** 1. tert Butyl cumyl peroxide

2. $CH_3-C(CH_3)_2-O-O-C(CH_3)_2-C_6H_5$

3. Trigonox — Noury

Grades T, TV–50

4. Grade T: Colourless liquid. M.W. 208, M.P. <17 °C.; S.G. 0.96.

Grade TV–50: White powder, 50 % active ingredient with silica filler. S.G. 1.2.

Vulc. for NR, synthetic rubbers, silicone rubber and polyolefins.

**23** 1. tert Butyl perbenzoate

2. $CH_3-C(CH_3)_2-O-O-C(=O)-C_6H_5$

3. Trigonox C — Noury

4. Clear slightly yellow liquid. M.W. 194; Flash Point 50 °C.; S.G. 1.04.
   Vulc. for silicone rubber.

**24**
1. Di-tert butyl peroxide
2. 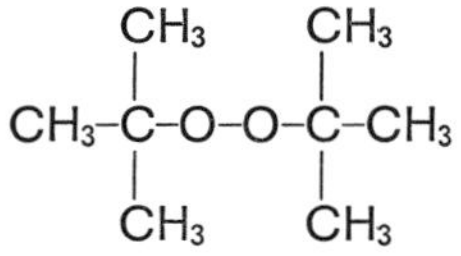 
$$
\begin{array}{ccccccc}
 & CH_3 & & & CH_3 & & \\
 & | & & & | & & \\
CH_3- & C & -O-O- & & C & -CH_3 \\
 & | & & & | & & \\
 & CH_3 & & & CH_3 & &
\end{array}
$$
3. — Wallace & Tiernan a
   Trigonox B Noury
4. Colourless, clear liquid. M.W. 146; B.P. 111 °C.; Flash Point 8 °C.; S.G. 0.793.
   a. B.P. 119 °C.
   Vulc. for natural- and synthetic rubber, silicone rubber and polyolefins.

**25**
1. 3,3-Di-tert butyl-peroxy-butanecarboxylic-n-butyl-ester
2. 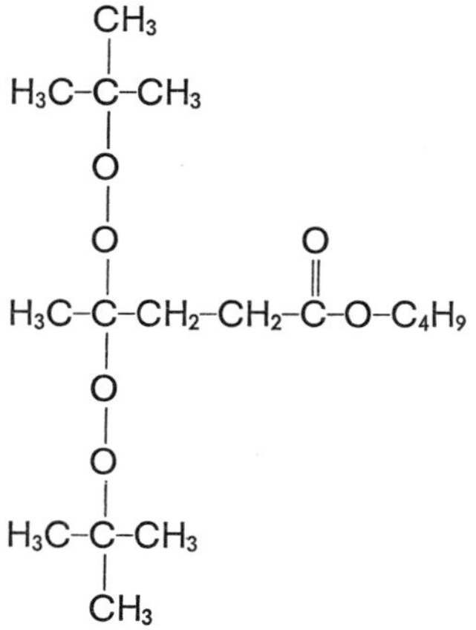 
$$
\begin{array}{lcl}
 & CH_3 & \\
 & | & \\
H_3C- & C & -CH_3 \\
 & | & \\
 & O & \\
 & | & \\
 & O & \qquad\quad O \\
 & | & \qquad\quad \| \\
H_3C- & C & -CH_2-CH_2-C-O-C_4H_9 \\
 & | & \\
 & O & \\
 & | & \\
 & O & \\
 & | & \\
H_3C- & C & -CH_3 \\
 & | & \\
 & CH_3 &
\end{array}
$$
3. Trigonox Noury
   Grades 17, 17/40.
4. Grade 17: Pale yellow liquid. M.W. 334; S.G. 0.95.
   Grade 17/40: White powder, 40% active ingredient with $CaCO_3$. S.G. 1.56.
   Vulc. for natural- and synthetic rubber, silicone rubber and polyolefins.

**26** 1. 1,1-Di-tert butyl-peroxy-3,3,5-trimethyl cyclohexane

2.

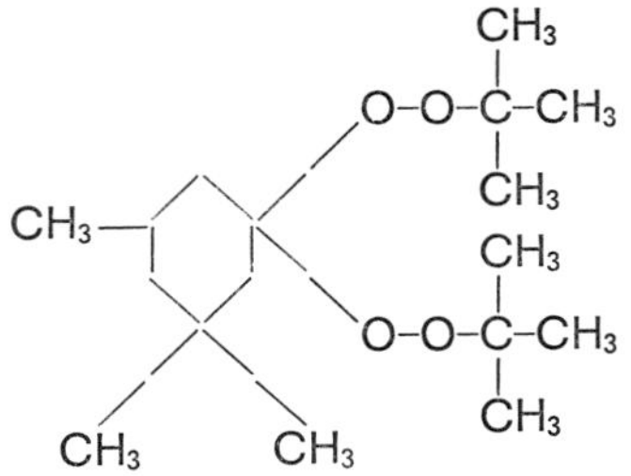


3. Luperco 231–XL — Wallace & Tiernan
   (40 % on inert filler)
   Luperco 231–XLP — Wallace & Tiernan
   (30 % paste)
   Luperox 231 — Wallace & Tiernan
   (93 % liquid)
   Percadox 29/40 — Noury
   (40 % on $CaCO_3$)
   Trigonox 29/40 — Noury

4. M.W. 302.44.
   Vulc. for natural- and synthetic rubber, and for silicone rubber.

**27** 1. Dicumyl peroxide

2.

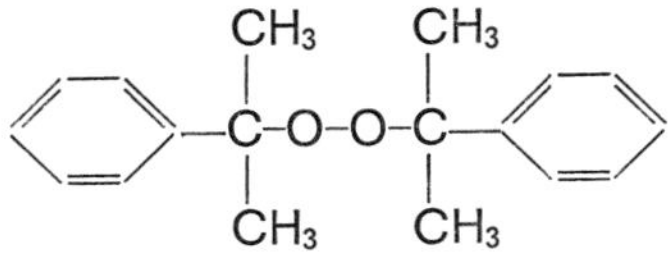


3. — — Wallace & Tiernan a
   Di-Cup — Hercules b
   Grades 40C, 40KE, R, T
   Perkadox — Noury c
   Grades BC 40, SB

4. M.W. 270; M.P. 42 °C.
   a. 95 % active ingredient on inert filler; 40 % active ingredient on inert filler.
   b. Grade 40C: White powder, 39–41 % active ingredient. S.G. 1.53.
      Grade 40KE: White powder, 39.5–41.5 % active ingredient. S.G. 1.607.
      Grade R: White granules, 96–100 % active ingredient. M.P. 38 °C.; S.G. 1.018.

Grade T: White crystalline powder, 90–93 % active ingredient. M.P. 30 °C.; S.G. 1.023.

c. Grade BC 40: White powder, 40 % active ingredient. S.G. 1.53.

Grade SB: White crytalline powder. M.P. 42 °C.; S.G. 1.02.

**28** 1. Di-(2,4-dichlorobenzoyl) peroxide

2. Cl–$C_6H_3$(Cl)–C(=O)–O–O–C(=O)–$C_6H_3$(Cl)–Cl

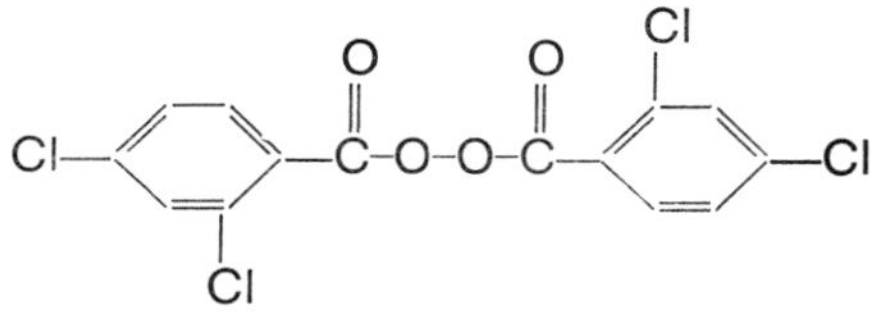

3. Cadox TS 50 — Nourychem a
Luperco — Wallace & Tiernan b
Grades CDB, CST.
Norox DBP–S–50 — Norac c
Perkadox — Noury d
Grades PDB 50, PDS 50, SD.
Siloprene Crosslinking Agent CL 40 — Bayer e

4. M.W. 380.0.
a. 50 % Paste in dibutylphthalate and silicone oil. S.G. 1.24.
b. Grade CDB: 50 % paste in dibutylphthalate.
Grade CST: 50 % paste in silicone oil.
c. 50 % Paste in polydimethylsiloxane 1000.
d. Grade PDB 50: 50 % paste in dibutylphthalate.
Grade SD: white powder, 95 % active ingredient.
e. 40 % Paste in silicone oil. S.G. 1.18.
Vulc. for silicone rubber.

**29** 1. 2,5-Dimethyl-2,5-(di-tert butyl-peroxy) hexane

2. $H_3C–C(CH_3)(O–O–C(CH_3)_3)–CH_2–CH_2–C(CH_3)(O–O–C(CH_3)_3)–CH_3$

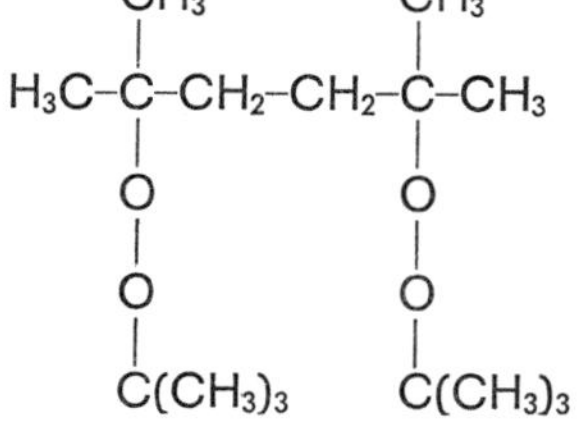

3. Luperco 101 XL — Lucidol a
   Luperco 101 XL — Pennwalt a
   Luperco 101 XL — Wallace & Tiernan a
   Lupersol 101 — Lucidol b
   Lupersol 101 — Pennwalt b
   Lupersol 101 — Wallace & Tiernan b
   Varox — Vanderbilt c
4. M.W. 290.45.
   a. White powder, 45–50 % active ingredient on inert carrier. S.G. ca. 1.50.
   b. Water-white liquid. B.P. 250 °C.; S.G. 0.87.
   c. Liquid and powder.

   Crosslinking agents for SBR, EPDM, PU, and silicone rubber.

**30**

1. 2,5-Dimethyl-2,5-(tert butyl-peroxy) hexyne-3
2.

$$\begin{array}{ccccc} & CH_3 & & CH_3 & \\ & | & & | & \\ CH_3- & C & -C\equiv C- & C & -CH_3 \\ & | & & | & \\ & O & & O & \\ & | & & | & \\ & O & & O & \\ & | & & | & \\ CH_3- & C & -CH_3 \quad CH_3- & C & -CH_3 \\ & | & & | & \\ & CH_3 & & CH_3 & \end{array}$$

3. Luperco 130 XL — Wallace & Tiernan a
   Luperco 130 XL — Lucidol a
   Luperox 130 — Wallace & Tiernan b
   Lupersol 130 — Lucidol b
4. a. White powder, 45 % active ingredient. Bulk density 0.481.
   b. Light-yellow liquid. M.W. 286.42; Flash Point 95 °C.; B.P. 243 °C.

   Crosslinking agents for EPR.

**31**

1. 4-Methyl-2,2-bis-(tert butyl-peroxy) pentane
2.

$$\begin{array}{l} CH_3-CH-CH_2-C-CH_3 \\ \quad\;\; | \qquad\quad\; / \;\; \backslash \\ \quad CH_3 \quad\;\; O \quad\; O \\ \qquad\qquad\;\; | \qquad | \\ \qquad\qquad\;\; O \quad\; O \\ \qquad\qquad\;\; | \qquad | \\ \qquad (CH_3)_3C \quad C(CH_3)_3 \end{array}$$

3. Luperco 144 XL — Wallace & Tiernan
4. Powder, 40 % active ingredient on inert carrier.
   Vulc. for EPR, EPDM.

**Thiuram-sulphides**

**32**
1. Dipentamethylene-thiuram disulphide
2. $(C_5H_{10}N.CS.S)_2$
3. Robac P.T.D. — Robinson a
   Robac P.T.D. 86 — Robinson b
   (containing extra sulphur)
4. Cream coloured powder. M.W. 320.6.
   a. M.P. 120 °C.
   b. M.P. 110 °C.
   Non-staining vulc. and acc. for latex (gloves) and for IIR (pharmaceutical closures).
   △ Acc. 75

**33**
1. Dipentamethylene-thiuram hexasulphide
2. $$\mathrm{CH_2}\langle\begin{smallmatrix}\mathrm{CH_2{-}CH_2}\\ \mathrm{CH_2{-}CH_2}\end{smallmatrix}\rangle\mathrm{N{-}\overset{S}{\overset{\|}{C}}{-}(S)_6{-}\overset{S}{\overset{\|}{C}}{-}N}\langle\begin{smallmatrix}\mathrm{CH_2{-}CH_2}\\ \mathrm{CH_2{-}CH_2}\end{smallmatrix}\rangle\mathrm{CH_2}$$
3. DPTT — Akron
   Sulfads — Vanderbilt
   Tetrone A — Du Pont
4. Light gray powder. M.W. 448; M.P. 110 °C.; S.G. 1.53.
   Vulc. for IR; non-staining acc. for NR, SBR, NBR, CR, IR, IIR, EPDM, CSM. Also for latex.
   △ Acc. 77

**34**
1. Dipentamethylene-thiuram-tetrasulphide
2. $(CH_2)_5N.CS.S_4.CS.N(CH_2)_5$
3. Accelerator 4P — Anchor
   DPTT — Hasselt
   Nocceler TRA — Ouchi Shinko
   Robac P.25 — Robinson
   Soxinol TRA — Sumitomo

4. Light yellow powder or rods. M.W. 384.69; M.P. 115 °C.; S.G. 1.50.
   Vulc. and ultra-acc. for CSM, IIR, EPDM, NBR, SBR, IR, CR, NR; act. for thiazole- and sulphenamide-type acc.
   △ Acc. 76
   △ Act. 7

**35**
1. Tetrabutyl-thiuram disulphide
2. $[(C_4H_9)_2N.CS.S-]_2$
3. Robac T.B.U.T. — Robinson
   Soxinol TBT — Sumitomo
   TBTS — Bozzetto
4. Brown liquid. M.W. 408.7; solidifies at ca. 20 °C.; S.G. 1.1. Insoluble in water.
   Vulc. for NR, SBR, NBR; non-staining acc. in combination with P.T.D. or T.M.T. for NR, SBR, NBR in sulphurless compounds; ret. for CR.
   △ Acc. 79
   △ Ret. 10

**36**
1. Tetraethyl-thiuram-disulphide
2. $(C_2H_5)_2N.CS.S.S.CS.N(C_2H_5)_2$
3. Accicure TET — Alkali
   Aceto TETD — Aceto
   Ancazide ET — Anchor
   Ethyl Thiram — Pennwalt
   Ethyl Thiurad — Monsanto
   Ethyl Tuads — Vanderbilt a
   Ethyl Tuex — Naugatuck SpA
   Ethyl Tuex — Uniroyal
   Etiurac — Ticino
   Eveite T — ACNA Montecatini
   Hermat TET — Dimitrova
   Nocceler TET — Ouchi Shinko
   Robac TET — Robinson
   Sanceler TET — Sanshin
   Soxinol TET — Sumitomo
   Superaccelerator 481 — Rhône Poulenc

| | |
|---|---|
| TETD | Hasselt |
| TETD | Prochim |
| TETS | Bozzetto |
| Thiuram E | Du Pont |
| Vondac TET | Vondelingenplaat |
| Vulcafor TET | ICI |
| Vulcafor TET | ICI (India) |

4. Grayish-white powders and pellets. M.W. 296.54; M.P. 71–73 °C.; S.G. 1.26.
   a. Powder and rods.
   Vulc. for sulphurless compounds; acc. for NR, SBR, NBR, IIR, IR, EPDM; act. for thiazole-, guanidine- and aldehyde-type acc.; stabilizer for Neoprene GN. Non-staining.
   △ Acc. 81
   △ Act. 8
   △ Antidegr. 96

**37**

1. Tetramethyl-thiuram disulphide
2. $(CH_3)_2N.CS.S.S.CS.N(CH_3)_2$
3. Accicure TMT — Alkali

| | |
|---|---|
| Accicure TMT | Alkali |
| Aceto TMTD | Aceto |
| Ancazide ME | Anchor |
| Cyuram DS | Cyanamid |
| Eveite 4MT | ACNA Montecatini |
| Hermat TMT | Dimitrova |
| Methyl Thiram (oiled and extruded) | Pennwalt |
| Methyl Tuads | Vanderbilt a |
| Metiurac | Ticino |
| Nocceler TT | Ouchi Shinko |
| RC Granulat TMTD (80 % TMTD, 20 % saturated hydrocarbons and dispersing agents) | Rhein Chemie |
| Robac TMT | Robinson |
| Sanceler TT | Sanshin |
| Soxinol TT | Sumitomo |
| Superaccelerator 501 | Rhône Poulenc |

| | |
|---|---|
| Thiurad | Monsanto |
| Thiuram 16 | Metallgesellschaft |
| Thiuram M | Du Pont |
| TMTD | Hasselt |
| TMTD | Akron |
| TMTD | Prochim |
| TMTS | Bozzetto |
| TMTS oleato (83 % TMTS, 17 % paraffin oil) | Bozzetto |
| Tuex | Naugatuck SpA |
| Tuex | Uniroyal |
| Vondac TMT | Vondelingenplaat |
| Vulcafor TMT | ICI |
| Vulcafor TMT | ICI (India) |
| Vulkacit Thiuram | Bayer |
| Vulkacit Thiuram C (surface coated) | Bayer |
| Vulkacit Thiuram GR (granules) | Bayer |

4. White to yellow powder. M.W. 240.44; M.P. 146–148 °C.; S.G. 1.3–1.4.
   a. Powder and rods.
   Non-staining ultra-acc. and vulc.; act. for thiazole- and sulphenamide-type acc.
   △ Acc. 83
   △ Act. 9

**38** 1. Tetraethyl-thiuram disulphide / tetramethyl-thiuram disulphide blend
2. —
3. Methyl Ethyl Tuads — Vanderbilt a
   Pennac TM — Pennwalt b
4. a. 50 : 50 mixture. White to-cream rods. M.P. 62 °C. min.; S.G. 1.32.
   b. Light buff powder. M.W. 257.5; S.G. 1.24.
   Vulc. and acc. for sulphurless or low-sulphur stocks of NR, SBR, EPDM.
   △ Acc. 84

## Carbamates

**39** 1. Hexamethylene-diaminecarbamate

2. $H_3\overset{\oplus}{N}-(CH_2)_6-N(H)-\overset{\ominus}{CO_2}$

3. Diak No. 1 — Du Pont
   Tecnocin B — ACNA Montecatini

4. White powder. M.W. 160; M.R. 152–155 °C.; S.G. 1.15.
   Vulc. for fluoroelastomers.

**40** 1. Selenium diethyldithiocarbamate

2. $[(C_2H_5)_2N.CS.S-]_4Se$

3. Ethyl Selenac — Vanderbilt
   Ethyl Seleram SA-66-1 (oiled) — Pennwalt
   Seleniame — Prochim
   Soxinol SE — Sumitomo

4. Yellow-orange powder. M.W. 672; M.R. 59–85 °C.; S.G. 1.32.
   Vulc. for NR, NBR, SBR; acc. for IIR; act. for thiazole-type acc.
   △ Acc. 57
   △ Act. 23

**41** 1. Selenium dimethyldithiocarbamate

2. $[(CH_3)_2N.CS.S-]_4Se$

3. Methyl Selenac — Vanderbilt

4. Yellow powder and rods. M.W. 559.78; M.R. 140–172 °C.; S.G. 1.58.
   Vulc. and acc. for NR, IIR, SBR, BR, IR.
   △ Acc. 58

## Miscellaneous, mixtures and undisclosed compositions

**42** 1. Dibenzoyl-p-quinonedioxime

2. $C_6H_5-CO-O-N=C_6H_4=N-O-CO-C_6H_5$

3. Dibenzo G-M-F — Uniroyal
   Dibenzo PQD — Conestoga
   Kencure Dibenzo Q.D.O. — Kenrich
   Vulnoc DGM — Ouchi Shinko

Vulnoc DGM paste — Ouchi Shinko
(50 % active ingredient)

4. Gray-brown powder. M.W. 346; S.G. 1.37; decomposition above 200 °C.
Vulc. for NR, SBR, IR, IIR.

**43**
1. dimeric 2,4-Toluylenediisocyanate
2. (structural formula: two tolyl rings, each with $CH_3$ and NCO, joined through N to a four-membered ring N–C(=O)–N–C(=O))
3. Desmodur TT — Bayer
4. White powder. M.W. 348; M.P. 145 °C. min.; S.G. 1.48.
Curing agent for urethane elastomers.

**44**
1. 4,4'-Dithiodimorpholine
syn. Dimorpholinyl disulphide
morpholine disulphide
2. $O(CH_2\text{-}CH_2)_2N\text{-}S\text{-}S\text{-}N(CH_2\text{-}CH_2)_2O$
3. Deovulc M — Grandel
Sulfasan R — Monsanto
Vanax A — Vanderbilt
Vondac DTDM — Vondelingenplaat
Vulnoc R — Ouchi Shinko
4. White to grey powder. M.W. 236.36; M.R. 115–127 °C.; S.G. 1.35. FDA appr.
Vulc. and acc. for NR, SBR, NBR, IIR, EPDM.
△ Acc. 85

**45**
1. N,N'-Di-thio-bis-(hexahydro-2H-azepinon-2)
2. $CH_2\text{-}CH_2\text{-}CO$ / $CH_2\text{-}CH_2\text{-}CH_2$ ring N–S–S–N $CO\text{-}CH_2\text{-}CH_2$ / $CH_2\text{-}CH_2\text{-}CH_2$ ring

3. Rhenocure S — Rhein Chemie
4. Cream powder or granules. M.W. 288; S.G. 1.3.
   Non-staining vulc. for natural and synthetic rubber.

**46**
1. Hydroquinone (treated)
2. $C_6H_4(OH)_2$
3. Diak No. 5 — Du Pont
4. Off-white powder. M.W. 110; M.P. 170 °C.; S.G. 1.3.
   Curing agent for Viton D-80.

**47**
1. Hydroquinone-dihydroxyethylether
2. $HO{-}CH_2{-}CH_2{-}O{-}C_6H_4{-}O{-}CH_2{-}CH_2{-}OH$
3. Crosslinking Agent 30/10 — Bayer
4. White crystalline powder. M.W. 198; M.P. 104 °C.; S.G. 1.34.
   Sec. crosslinking agent for urethane elastomers in combination with Desmodur TT (dimeric 2,4-toluylenediisocyanate).

**48**
1. organic Lead salts
2. —
3. Desmorapid 600 — Bayer
   Desmorapid DA — Bayer
4. Yellow powder. S.G. 2.
   Crosslinking agents for urethane elastomers.

**49**
1. Poly-p-dinitrosobenzene
2. $[O{=}N{-}C_6H_4{-}N{=}O]_n$
3. Vulnoc DNB — Ouchi Shinko
4. M.W. $(136.11)_n$
   Vulc. for IIR.

**50** 1. Poly-2,2,4-trimethyl-1,2-dihydroquinoline

2. [structural formula: repeating unit of 2,2,4-trimethyl-1,2-dihydroquinoline (NH, $CH_3$, $CH_3$, CH, C–$CH_3$)]$_n$

3.

| | |
|---|---|
| Aceto POD | Aceto |
| Agerite AK (tradename outside U.K. Antioxidant 184) | Anchor |
| Agerite Resin D | Vanderbilt |
| Anox HB | Bozzetto |
| Antigene RD | Sumitomo |
| Antioxidant HS | Bayer a |
| Flectol H | Monsanto |
| Nocrac 224 | Ouchi Shinko |
| Pennox HR | Pennwalt |
| Permanax 45 | Rhône Poulenc |

4. Buff powder or flakes. M.P. 100–120 °C. (a. M.P. 75 °C. min.); S.G. 1.08.
Vulc. for CR; semi-staining oxi, heat, inh for NR, NBR, BR, IR, CR. Also for latex.
△ Antidegr. 95

**51** 1. p-Quinone-dioxime

2. HO–N=$C_6H_4$=N–OH

3.

| | |
|---|---|
| Curing Agent CDO 50 | Bayer |
| G–M–F | Uniroyal |
| Kencure QDO | Kenrich |
| PQD | Conestoga |
| Vulcafor BQ | ICI † |
| Vulnoc GM | Ouchi Shinko |
| Vulnoc GM paste (50 % active ingredient) | Ouchi Shinko |

4. Dark-brown powder. M.W. 138.13; M.P. 215 °C.; S.G. 1.97.
Vulc. for NR, SBR, IIR and polysulphiderubber.

**52**
1. Tetrachloro-p-benzoquinone
2. O:$C_6Cl_4$:O–1,4
3. Vulklor — Uniroyal
4. Yellow powder. M.W. 244; M.P. 290 °C.; S.G. 1.97.
   Non-staining vulc. for NR, NBR, IIR, SBR. (sulphurless vulcanization)

**53**
1. Zincchloride - benzothiazyl disulphide complex
2. —
3. Caytur 4 — Du Pont
   (formerly LD–755)
4. Yellow powder. M.P. 235 °C.; S.G. 1.85.
   Crosslinking agent for sulphur-curable urethane-elastomers.

**54**
1. Composition undisclosed
2. —
3. Diak No. 4 — Du Pont a
   Kenmix — Kenrich b
   Synform C 1000 — Anchor c
   Trigonox X–17/40 — Noury d
4. a. White powder. M.R. 145–155 °C.; S.G. 1.23.
   Curing agent for fluoro-elestomers.
   b. △ Acc. 141
   △ Act. 44
   △ Antidegr. 125
   c. Orange-yellow lumps. M.P. 65 °C.
   Curing agent for IIR.
   d. Powder.
   General purpose.

# Subject index

**Abbreviations**

| | | | |
|---|---|---|---|
| Accelerators | Acc. | Co-agents | Ca. |
| Activators | Act. | Peptizers | Pept. |
| Antidegradants | Antidegr. | Retarders | Ret. |
| Blowing agents | Ba. | Vulcanizing agents | Vulc. |

# Trade names index

# Contents